Rainfed Altepetl

Modeling institutional and subsistence agriculture in ancient Tepeaca, Mexico

Aurelio López Corral

Archaeopress Pre-Columbian Archaeology 3

Archaeopress
Gordon House
276 Banbury Road
Oxford OX2 7ED

www.archaeopress.com

ISBN 978 1 78491 040 2
ISBN 978 1 78491 041 9 (e-Pdf)

© Archaeopress and A López Corral 2014

All rights reserved. No part of this book may be reproduced, stored in retrieval system,
or transmitted, in any form or by any means, electronic, mechanical, photocopying or otherwise,
without the prior written permission of the copyright owners.

Printed in England by CMP (UK) Ltd

This book is available direct from Archaeopress or from our website www.archaeopress.com

Table of Contents

Chapter 1
Introduction 1
   The goals of this work 1
   Late Postclassic Tepeaca agriculture: a dualistic model 2
      Studying agricultural production variability at the household and regional level 3
      Regional agricultural production variation in Tepeaca: an ethnographic work 5
      Chapter organization and content 7

Chapter 2
Agriculture And Theory 8
   Subsistence agriculture 8
   Institutional agriculture 12
   Environmental and cultural variables that affect agricultural production 13
      Water availability 13
      Soils and sediments 13
      Climate variability 14
   Agro-ecological variables that affect agricultural production 15
      Plant Sowing Densities and Farming Strategies 15
   Cultural factors that affect agricultural production 16
   Mesoamerican agriculture: rainfed dependent and artificially supplied water system 18
   Agricultural food shortage 19

Chapter 3
The Natural Setting 21
   The Tepeaca valley region 24
   The Llanos de San Juan 27
   The Puebla-Tlaxcala valley 28

Chapter 4
Regional History and the Tepeaca *Altepetl* 30
   Tepeaca and the Puebla-Tlaxcala valley during the Postclassic 30
      Regional history and the origins of the Tepeaca altepetl 30
      Social structure in Tepeaca and the Puebla-Tlaxcala valley in the 16th century AD 33

Chapter 5
Traditional Agriculture in the Study Region 36
   Stages and work in the agricultural cycle 36
      Field preparation 36
      Furrowing and sowing 37
      Animal manure fertilization 39
      Seed selection 39
      Re-sowing 40
      Primera, segunda and tercera labor (first, second and third labor) 40
      Second weeding 40
      Chemical fertilization 40
      Maize stalk upper portion removal 41
      Harvest 41
      Storage 41
   Other aspects of local agriculture 42
      The use of High Yield Varieties (HYV's) 42
   Final remarks regarding modern agriculture in the study region 43

Chapter 6
Agricultural Production for the Year 2009: the Ethnographic Survey 45
   Part one: the household survey 46
      Methodology 46
      Measuring maize production 48
      Problems using weight measurement 48

Volume as an alternative option for determining production	49
Length of the ear and weight of the dry kernel	50
Length of the cob and weight of the dry kernel	50
Calculation of maize production per hectare	50
Initial plant densities, survival plant densities and total productivity	52
Result	53
Production from the surveyed field	54
Soils and production	54
Rain patterns and the 2009 canícula drought	57
Other negative phenomenon	60
Work inputs to field	63
Types of rain	63
Second part: regional agricultural productivity in the Tepeaca region	63
The 2009 maize production in the study region	66
Methodology	67
Maize yields in the study region according to land class	67
The Tepeaca valley	70
The Puebla-Tlaxcala valley	70
The Llanos de San Juan	70
Maize yields according to municipio	70
Differential sowing and harvest within the region	72

Chapter 7
From Prehispanic *Macehualli* to Colonial *Terrazgueros* 75
The prehispanic *macehualli* 76
Chichimec conquests in the Cuauhtinchan-Tepeaca region 76
Colonial *macehualli* and *terrazguero* 78

Chapter 8
Agricultural Productivity and Tribute in 16th Century AD Tepeaca 80
Land allotment and agricultural tribute in Early Colonial Tepeaca 80
    Land tenure in Early Colonial Tepeaca 80
    Types of length measures in Tepeaca 81
    The braza and the nehuitzantli 81
    The indigenous rod or tlalquahuitl 83
    The size of agricultural plot 84
Production capacity at the subsistence and institutional agriculture level 84
Institutional agriculture production 85
Subsistence agriculture production 88
    Maize production within macehualli/terrazguero households 88
Buffering strategies against climatic variability 90
    Field dispersion 90
    Agricultural intensification 92
    The marketplace 92
    Summary 96

Chapter 9
Conclusions and Directions for Future Research 97
Modeling agricultural productivity in ancient Tepeaca 97
Identifying buffering strategies against cyclical food shortfall 98
The dual agricultural economic structure of the Tepeaca *altepetl* 100
Directions for future research 101

References 103

Appendix
2009 Ethnographic Survey: Field Registers 115

Chapter 1
Introduction

The goals of this work

This study analyzes the impact of climate variability and human management strategies on the Late Postclassic (AD 1325-1521) and Early Colonial (16th century AD) agricultural systems in the Tepeaca Region, Puebla, Mexico. The research examines the scale of crop production at the subsistence and institutional agricultural levels and the role of commoner rural populations within the prevalent tributary economic system.

This work has three objectives: (1) to model agricultural productivity at the household and regional levels, (2) identify the buffering strategies developed by Tepeaca's populations against cyclical food shortfalls, and (3) establish a model for the agricultural and economic structure of the Tepeaca *altepetl* or state-level polity. Crop production in Tepeaca depended primarily on rainfall as the major source for water. Therefore, unpredictable and variable climatic conditions resulted in a low and unstable production potential. Other factors also constrained the productive capacity of local agricultural systems. These included the limited labor force of households, the simple agricultural technology and the prevailing land tenure systems. These factors had a profound effect on the agrarian structure of indigenous communities and on the tributary demands that political entities could levy on the peasant majority.

The model for the agricultural and economic structure of the prehispanic Tepeaca *altepetl* considers that the agricultural systems were arranged dualistically, similar to other aspects of the prehispanic culture. Two independent types of agriculture existed alongside one another each with a very different focus. On one level, there was subsistence agriculture characterized by a low-level production capacity and geared towards food production intended for auto-consumption within commoner peasant households. The other was institutional agriculture, which dealt largely with production for the support and finance of political institutions that included the nobility, military, theocratic, and bureaucratic sectors of the community.

The Tepeaca Region is located in the central portion of the State of Puebla, Mexico (Figure 1). It borders several important cultural areas of the central Mexican highlands like the Puebla-Tlaxcala Valley to the west, the Tehuacan region to the southeast and the Llanos de San Juan to the North (Figure 2). Within this macro-region, Formative communities flourished and became the basis for the subsequent development of complex agrarian societies and the dense urban settlements of the Classic and Postclassic

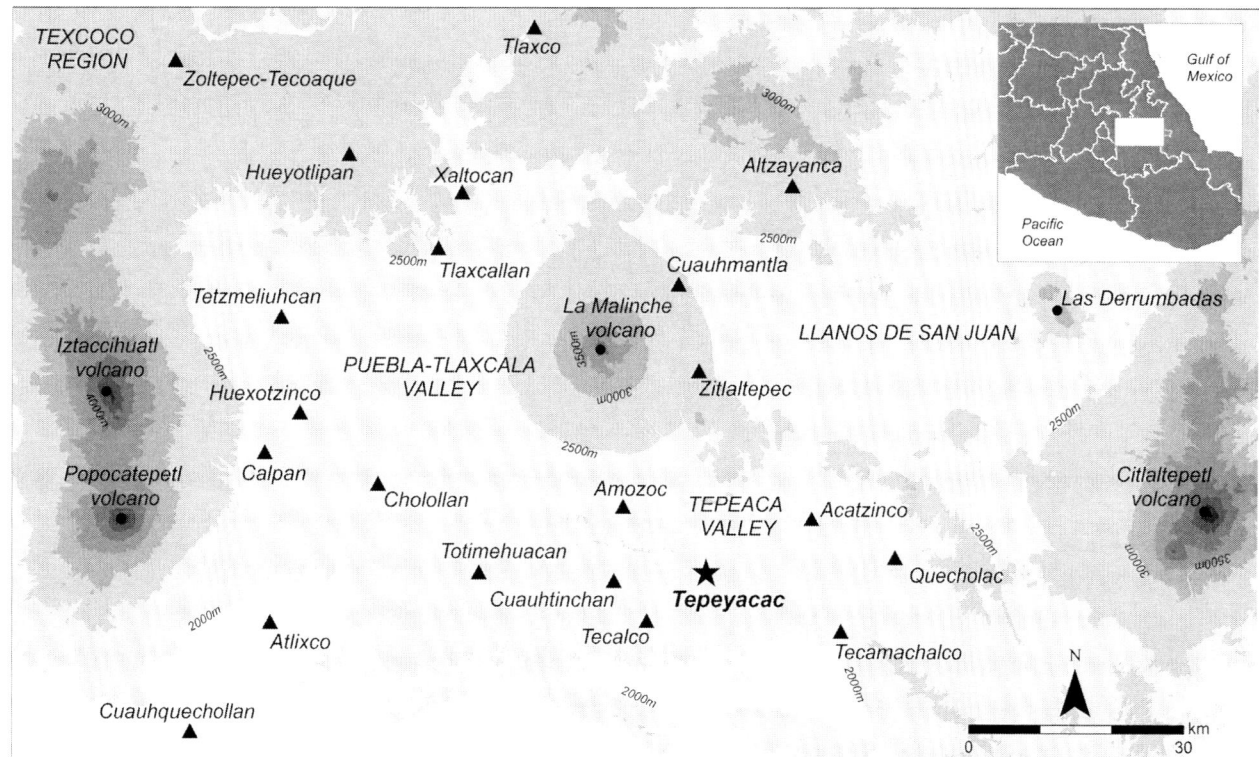

FIGURE 1. MAP SHOWING THE STUDY REGION AND VARIOUS POSTCLASSIC SETTLEMENTS WITHIN THE STATES OF PUEBLA AND TLAXCALA.

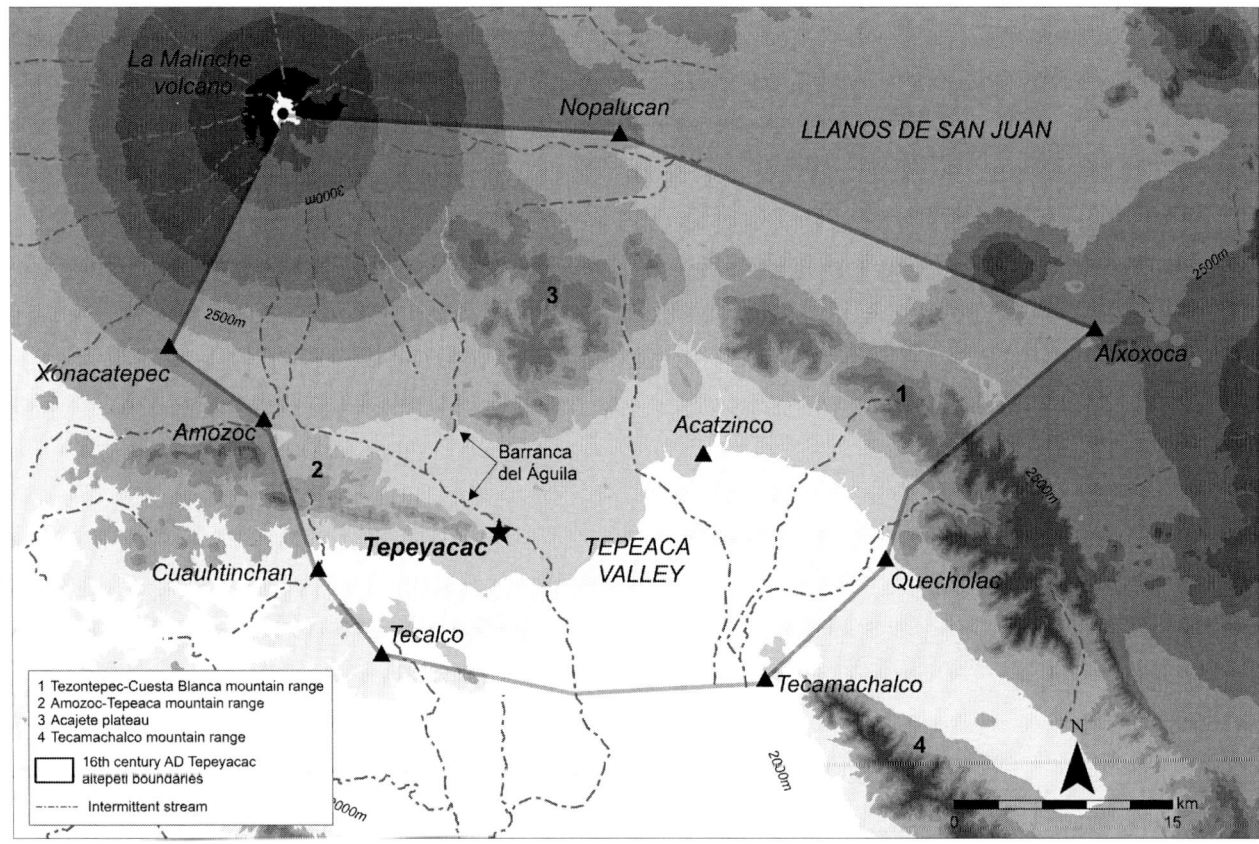

FIGURE 2. MAP SHOWING THE TERRITORIAL BOUNDARIES OF THE TEPEACA ALTEPETL DURING THE 16TH CENTURY AD.

periods (Castanzo 2002; Castanzo and Sheehy 2004: Fargher 2007; García Cook 1981, 1985; García Cook and Merino 1986; Hernandez Xolocotzi 1965; Kirchhoff et al. 1976; Martínez 1984b; Merino and García Cook 1998; Plunket 1990; Plunket and Uruñuela 1998, 2005; Prem 1978; Sheehy et al. 1997).

Tepeyacac Tlayhctic, known today as the town of Tepeaca, was an important *altepetl* throughout the Late Postclassic (See Chapter 2). According to the *Historia Tolteca Chichimeca* (Kirchhoff et al. 1976: [319, 320]) the town was first founded around AD 1178 by a migrant group known as the colhuaque. Four years later in AD 1182 a second wave of immigrants known as the tepeyacatlaca also settled in the area. Tepeaca formed part of the Cuauhtinchan *altepetl*, the region's ruling entity composed of a multiethnic population that controlled a vast territory during the Middle and Late Postclassic (AD 1100-1521). Late Postclassic conflicts between both settlements led to the defeat and expulsion of the Cuauhtinchan rulers in AD 1457 This was followed by two years of turmoil when no community had complete control over the region, and seven more with the occupation of the Cuauhtinchan lands by Tepeaca.

Tepeaca was located on an important communication corridor connecting the Basin of Mexico to the West and Gulf Coast lowlands to the east and the Valley of Oaxaca to the southeast. This strategic location increased Tepeaca's commercial importance, but it also caught the attention of the imperialistic interests of the Mexica Empire who conquered it in AD 1466 under the rule of Moteuhcuzoma Xocoyotzin and the leadership of Axayacatl. This event brought about a rearrangement of the region's territorial boundaries, in which Tepeaca was given its own domain with tributary populations. These boundaries lasted until the Spanish arrival in the first part of the 16th century AD.

Another result of Tepeaca's conquest was that the Mexica ordered a marketplace be established in the town where their merchants could be hosted and where substantial amounts of goods would be made available for trade. The Tepeaca marketplace became so renowned that it survived the Colonial Period and continued up to modern times. Within the establishment of new boundaries, two other important communities Acatzingo and Oztoticpac where inserted into the political tributary system of Tepeaca. The Tepeaca *altepetl* shared boundaries with five other major political entities of its time: Cuauhtinchan, Tecalco, Quecholac, Tecamachalco and Tlaxcala. Nonetheless, historical sources show that the region was continuously embedded in conflicts and alliances between the various *altepemeh*, mainly because of territorial disputes and for the control of land and labor (Dyckerhoff 1978).

Late Postclassic Tepeaca agriculture: a dualistic model

Agriculture was the basis for the development of Mesoamerican complex societies. Yet, variability in production substantially influenced the economic structure of indigenous *altepemeh*, especially their corresponding tributary systems. In Tepeaca, agriculture production can be divided into two different strategies that had different goals and were performed by different sectors of the society (more specific characteristics are described in detail in Chapter 4). The non-elite or commoner sector of the society was engaged primarily in subsistence agriculture. This sector was controlled by independent peasant households and was destined mainly for their auto-consumption needs and securing access to other basic everyday products. In this type of agriculture, yields probably occurred at a low-level, just enough to procure food for the family's annual caloric requirements. The relatively small labor pool, the reliance primarily on rainfall for their crops, and the high risk of crop failure due to both environmental and human management variables restricted household agricultural yields. If there was a possibility for producing surplus above household consumption needs, it would most likely have been placed in storage or used in exchange items for other non-perishable products, a strategy that works well for coping with inter-annual and seasonal environmental unpredictability and localized regional food shortages (Halstead and O'Shea 1989).

Within the domestic economy, the production of crafts and utilitarian products was commonly carried out alongside food production (Hirth 2007, 2009). Those items produced solely for use within the household can be included alongside the subsistence agriculture because they were not destined for Institutional consumption outside the household. Probably, these types of goods were exchanged in the local marketplace for other items, an approach that would have been especially important during times of food shortages.

The second type of agricultural strategy is institutional agriculture. It deals mainly with the management and organization of agricultural systems destined for the support and finance of the political, religious, and other social institutions and the support of elite families. Although agriculture is generally performed by the peasant or commoner sectors of society, the goal of the political apparatus is to control the most productive lands and related technological advancements. Doing so, allows the political sectors to generate large staple food surpluses, or to cultivate special crops employed in the manufacture of wealth items (e.g., cotton for textile weaving or cacao as an elite restricted beverage). The production of staples destined for political propaganda like the celebration of rituals or festivities in the community also falls in this category. These events can be promoted and performed by the ruling institution of the society, as well as governmental or bureaucratic institutions from the non-elite sectors.

In prehispanic Tepeaca, the subsistence and the institutional agricultural strategies appear to have run side by side with each other. Tributary demands may have centered mostly on the production of wealth items (e.g., crafts), labor service within the royal houses and palaces, and community services such as the tequitl or tlacalaquilli. Although food production was an essential part of the tribute system and destined for the support of local elites and other political institutions (e.g., cleric, military, attached specialists), there were clear distinctions as to the distribution of work and the destination of the production. For Tepeaca, the lands of the political institutions were distributed among several peasant work groups. Each group worked only a fraction of land from a noble's estate aside from that needed for their auto-consumption needs. The size of the area they worked for the elite probably was around of 0.17ha and only represented around 1/5 to 1/7 of the area that each household worked annually. These tribute lands were not part of the household subsistence base of peasant households and thus did not interfere in any significant way beyond the labor invested with their respective food production. This scheme allowed the local peasants to pay their obligations and at the same time they worked enough land to sustain them. It also permitted them engage in other non-agricultural work, such as craft production or animal breeding, as supplemental economic strategies.

If elite or nobles would have interfered with the production base of commoner households by extracting tribute from their subsistence base, this would have inevitably undermined their reproduction and affected the goals of the political institutions by devastating its productive base. It would also have diminished the military resources available to the elite because the commoner sectors of society comprised the bulk of the manpower when conflict arose. This is an important consideration because war and conquest were a recurring phenomenon during Late Postclassic times.

Because subsistence agriculture was a separate sector it continued intact after the disappearance of indigenous institutional agriculture shortly after the Spanish conquest. Thus, the dualistic agricultural model is appropriate for explaining the persistence of the subsistence agriculture and its transformation into what is now known as traditional agriculture. Its relative simplicity and resistance to change allow it to survive the deep cultural and economic transformations that occurred during the Colonial Period. In contrast to subsistence agriculture, indigenous institutional agriculture crumbled and fell under the military, economic, ideological and religious impositions set by the Spanish after the Conquest. The dualistic agriculture model proposed here serves as a useful analytical view to understand the changing processes of the Late Postclassic and Early colonial agricultural systems of the Tepeaca *altepetl*. By extension, the model can also be applied to other contemporary communities of the central Mesoamerican highlands.

Studying agricultural production variability at the household and regional level

Agricultural systems by nature vary in their productive capacity. Environmental uncertainties, climatic variability, ecological settings, pests, diseases and differential managerial behavior are factors that determine the degree of success or failure of agricultural production (McGregor and Nieuwolt 1998). Modeling natural phenomena and their impact on food production systems of ancient societies has been a central part of archaeological research (Dincauze 2000). At the same time, human responses to environmental changes or human induced changes on the natural environment comprise an extensive area of research within anthropological studies (Goudie 2001; Green 1980; Lailand and Brown 2006; Redman 1999; Smith 2007).

Although archaeologists acknowledge the complex nature of agrarian systems and the variables involved in food production, the central tendency of most researchers when reconstructing ancient agricultural systems and their production capacity has been to emphasize simplicity over complexity (Halstead and O'Shea 1989: 2). What is disregarded in the process is the volatile nature of staple food production and its economic implications for household subsistence strategies and political surplus extraction. Therefore, a major issue discussed in this work is the advantage of using variable production estimations in addition to simple averages for reconstructing prehispanic economies. Variable production estimations permit a better understanding of the structure of indigenous agricultural economic systems. Contradictory as it may seem, the unpredictable nature of production allows us to establish with more certainty the capacity for surplus generation at the household level and its alienation on part of political institutions. Also, the study of agricultural production variability provides a more inclusive way of comprehending the buffering strategies at work in societies for coping with seasonal and inter-annual environmental uncertainties that generate food stresses among populations.

Generally, archaeological research has employed average agricultural production values when reconstructing socio-demographic processes and the development of complex agricultural systems over time. Examples include the reliance on domesticates as a major food source among food producers or agrarian societies (e.g., Smith 2001) or calculations of carrying capacity in different environmental zones within a region (e.g., Nicholas 1989; Sanders *et al.* 1979). I am not saying that we should not employ average values in research. On the contrary, they can be very useful if the research goals are directed towards understanding agricultural development processes at a broad scale. However, if the interest of the investigator is to detect and analyze divergences from the mean values and the stability of a given system, then we should look deeper into the broader fluctuations in agricultural output. This is especially true when analyzing procurement strategies at the subsistence level. At the household level, many agricultural tasks are responses to environmental uncertainties and minor climatic seasonal fluctuations. Cultural responses concerning when and where tasks are to be performed can vary widely between households due to collective and individual decision-making. Timing of sowing, construction and maintenance of agricultural features, planting patterns, weeding, crop fertilization and transplanting are some of the tasks that need to be fulfilled with anticipation and as efficiently as possible. Yet, deciding on the proper time to perform these tasks depend on each farmer's choice and needs, requirements, characteristics and particular setting of his fields within the landscape. In the final analysis, these individual decisions are very important for differential patterns in yields between households.

When archaeologists apply mean values to overall agricultural production within a settlement or a region it obscures the natural characteristics of agricultural production variation. Standardization can lead to the oversimplified view that a good or bad year's harvest will affect equally everyone within the community. An extreme view would be to consider that households are constantly able to amass substantial amounts of surplus production during 'normal' to 'good years' and that only under prolonged droughts episodes might they drastically become impoverished or die. In reality, this is not so. Each year a sector of the population produces well and others do badly. It is only during exceptionally good or bad years, or a series of them, that the majority of the population will be benefited or affected more uniformly, and even then, some sectors will deviate considerably from the year's mean values.

At the institutional level, the standardization of production estimates has been used to establish the level of food surplus generated and its extraction via the tributary systems. Models that prefer the 'top-down' views have centered on analyzing the control of production over intensive agricultural systems involving artificial water supplies (e.g., chinampas, drained fields, irrigation) (e.g., Billman 2002; Mountjoy and Peterson 1973; Parsons 1991, 1992) or the production and control of special wealth staples such as cotton (e.g., D'Altroy and Earle 1985; Smith and Hirth 1988). At other times, tribute in staple food products has been taken as being homogeneous across all sectors of the community and do not consider the dualistic nature of late Postclassic agricultural production. As I mentioned before, it is probable that staple extraction came mainly from wealth items and not through agricultural tribute imposed on the households. There were certain demands on agricultural production, but it seems they were not placed on the food resource base of households. Rather, demands focused on the labor force of peasants, one that was destined for working the public and institutional fields.

At least for part of central Puebla and Tlaxcala, Late Postclassic institutional agriculture involved marked differences in access to the means of production and the distribution of land. Like Chalco (Jalpa 2009), Cuauhtinchan (Reyes 1988a), Huexotzingo (Brito 2008), Morelos (Smith 1993), and Otumba (Evans 2001) agricultural land in Tepeaca was unequally distributed (Martínez 1984b) and virtually all land was under the control of the local political institutions. Individual commoner households only possessed or worked fairly small plots. Great amounts of food were accumulated by the *teccalli* system, named *tlahtocayo* in the case of

Tepeaca, to support its noble elite. For nobles, having large land holdings located in different environmental settings and dispersed in several regions allowed them to have a more stable economic system by averaging overall agricultural production through the dispersal of plots across the landscape.

As I point out later in this work, the unequal distribution of land and the volatile nature of agricultural production generated a pattern in which most of the risk of crop failure was concentrated within the household realm. The bulk of local landless peasants were smallholders with their fields concentrated in relatively small areas. This pattern of land use made them susceptible to highly variable annual production losses due to extreme regional and local climatic fluctuations. Elites generated both greater and more stable quantities of staple products because their large landholdings were spread over larger areas, thus avoiding climatic variability and crop failure much more efficiently. Land control also allowed nobles to take control over landless peasant labor and tribute. This mainly took the form of the payment in wealth items and personal services rather than on their agricultural food base.

Regional agricultural production variation in Tepeaca: an ethnographic work

An important aspect of the dualistic agricultural model was to obtain information regarding production variability at the household and regional levels through time and space. By examining the configuration of modern smallholder's agricultural fields, subject to strong environmental variability and differential human management strategies under rainfed conditions, it is possible to reconstruct the range over which ancient agricultural production fluctuated. Also, further analogical inferences can be made regarding the relationship between food production at the domestic level and the involvement —or detachment— of the political institutions on their food production base.

Unfortunately, it is difficult to study the variability in ancient regional agricultural production solely using archaeological data or historical sources. On the one hand, archaeological work has centered on the reconstruction of ancient climatic regimes and changes, but these have involved mainly long-term changes expressed in terms of centuries or millennia. New studies in Mesoamerica have shed light on seasonal climatic variability, such as the use of tree ring analysis and sediments from lakes and lacustrine deposits (e.g., Leyden *et al.* 1996; Nichols 2009: 160-161, Pétrequin 1994; Stahle *et al.* 2011; Therrell *et al.* 2006), but this is restrained by the availability of a long data sequence from lacustrine deposits and the preservation of ancient wood samples. On the other hand, historical texts do not register in full detail fine scale climatic events nor do we have data on regional or local variation in production within a region in any given year. We have some information available on large-scale agricultural catastrophes that were recorded in several indigenous historical accounts during the Early Colonial period like the Anales de Tecamachalco (Solís 1992), the Códice Kingsborough (Valle 1992), Fray Diego Duran's (2006 [1579]) *Historia de las Indias de la Nueva España e Islas de Tierra firme*, and Chimalpahin's (1998) *ocho relaciones de Culhuacan*. However, the data are sparse and does not deal with the details (see also García *et al.* 2003).

An alternative approach is to use modern data on climatic variability and agricultural production distribution in order to approximate ancient patterns. The logic is that, although climate has changed in several occasions during human occupation in the central highlands, climate variability should have prevailed in ancient times as it still occurs today (Halstead and O'Shea 1989). Even if local agriculture systems might have changed for better or worse in any given period, environmental unpredictability would have been an important constraint for agricultural systems, especially under conditions of low-level technological development and rainfall systems. Still, under the best possible scenario, unexpected climatic phenomena can have substantial economic implications for peasant households and agrarian communities.

Hence, my research employed an ethnographic study that focused on registering variation in maize yields within the Tepeaca region. Its main goal was to establish the effects of micro-climatic fluctuations, essentially rainfall variability, on regional maize production at the household level. I wanted to detect how crop yields varied within a region and how this could be correlated with climatic variability, individual land management strategies and other environmental and cultural quotidian circumstances. The primary goal was to observe how far yields could diverge from the overall average. Households struggle against the effects of recurrent seasonal unpredictable climatic events because they strongly affect their ability to survive and reproduce.

Unfortunately, short maize stalks and fields invaded by weeds characterized most of the Tepeaca region landscape during autumn of 2009 when I initiated my study (Figure 3). 2009 was an unusual agricultural year. Rainfall was scarce during 2009 resulting in large patches of abandoned agricultural plots. The sharp climatic contrasts of drought and flood sharply affected the plant's development stages and resulted in poor crop stands. At the beginning of the season, peasants anticipated an excellent agricultural year because generous rains appeared early suggesting a water-plentiful year. Most people initiated their agricultural labors between late April and early May, which represented a good head start, and went on without any major problem during the first month. Surprisingly, conditions changed quickly and precipitation stopped in June ushering in the dry canícula period,[i] an intraestival drought, which lasted up to three months in some regions.

---

[i] Mexican peasants say that the canícula can enter either as a humid period or, as in the 2009 year, with windy and dry conditions.

FIGURE 3. FIELD INVADED BY WEEDS WITH SHORT DRIED MAIZE STALKS

FIGURE 4. CONTRAST BETWEEN TWO ADJACENT MAIZE STANDS SOWN IN MAY OF 2009 NEAR QUECHOLAC, PUEBLA

According to the State of Puebla statistic data, this extended dry period affected approximately 90% of maize production in the region, 60% of which resulted in total failure. In some areas a few agricultural stands survived and people were able to harvest some maize, about 20% to 40% of the field's average production capabilities. The greatest regional effect was in the eastern part of the valley, in communities such as Acatzingo, Quecholac, Tepeaca, and Tenango where virtually all maize stands were lost. The western part of the valley of Puebla-Tlaxcala, which lies outside the study region and comprises towns such as Acatepec, Nealtican and Huejotzingo, was also affected by the drought, but there the rains ceased for only 30-45 days. In towns like Cholula, although overall precipitation was below historical averages, it nonetheless was stable and continuous enough to permit good production. A more extreme situation occurred in the southern parts of the region, in the towns of Atoyatempan, Cuauhtinchan, Tecamachalco and Tecali, where virtually all farmers had total losses. Once the dry period ended in late August, intense rain storms again ensued. However, the damage was already done: the crucial moment of maize pollination had long passed and bean crops died also out due to water deficiency. The second period of precipitation lasted through mid-October creating floods in some areas and accelerated the rotting of lifeless plants left in the fields.

Usually, the average annual rainfall is what is recorded for any particular year and throughout several years. To the researcher interested in collecting mean values, rather than the details of data variation, it would appear that water precipitation in Tepeaca for during 2009 was plentiful and representative of a good productive agricultural cycle. However, the problem was that precipitation fell unevenly. Massive water storms struck at the start and end of the rainy season. In the middle of the growing season, an important drought prevailed which devastated crop stands.

Local peasants could do little to cope with this unpredictable event. The out-of-the-ordinary year resulted in food scarcity in many towns —mainly in terms of maize, the dominant staple among Mexican smallholders. This precipitated a widespread loss of monetary income to families due to human energy invested in agricultural tasks and the cost of manure and fertilization products. Of course, not all production was lost. Those areas with deep-water well irrigation systems managed to produce at least two to three metric tons of maize. Yet, even these irrigated fields did not escaped the hazardous dry period. In Tepeaca, like in the Basin of Mexico (Sanders *et al.* 1979: 252), irrigation is commonly used to facilitate early sowing and to get a head start on the rainy season. When this is done, however, it is expected that during the rest of year the rains will provide the majority of the necessary water for plant development. What happened in 2009 is that the lack of rain resulted in crops producing well below what is considered an average to good harvest (up to six or eight tons/ha). Nonetheless, irrigated fields did produce higher yields than the rainfed ones, and the contrast between adjacent irrigated and rainfall fields was substantial (Figure 4).

This clearly showed that the main problem in the Tepeaca region is the lack of permanent water sources or substantial springs that can supply water artificially to agricultural plots. In the area, agriculture is risky and prone to crop failure. Many areas within the region have soils considered adequate or good for agriculture. However, without substantial water resources and a constant and correctly timed supply of water, small annual fluctuations in the weather cannot be adequately buffered. In areas with poor soils, such as the southern sections of the Tepeaca valley, negative conditions are intensified. Nevertheless, while these tribulations can cause severe economic deteriorations to local populations in any given year, farmers are not disheartened and continue to cultivate their plots in the next cycle, hoping for a good harvest at the end of each year. For them, it is a cultural tradition, a means of survival, and an enduring life style that has passed from generation to generation. It has valiantly resisted not only unpredictable seasonal environmental conditions, but also the uncertainties of modern economic systems and volatile governmental policies.

Like today, the Tepeaca region lacked permanent streams and rivers during the Late Postclassic and Early Colonial periods. The historic information suggests that rainfall agriculture dominated in the past as it does today. Rural populations had to depend on the unreliable seasonal rains for crop cultivation. Under such difficult conditions, one wonders how this type of agriculture was able to produce food for auto-consumption and a surplus that could mitigate inter-annual climatic variability.

In general, archaeology has attempted to reconstruct agricultural production capacity for ancient systems and their technological advancements. Yet, we do not clearly understand enough about the variation in production capacity and the critical division between surplus generation and food shortfalls among rural peasant households. It is very important to understand the production capacity of agricultural systems based on rainfall conditions, because of their susceptibility to fluctuations in the timing and amount of rainfall. Micro- and macro-seasonal climatic fluctuations were one of the major constraints for social development among Mesoamerican populations, especially inconsistent rains in time and space. At one extreme, droughts can produce massive crop failure and food shortfalls, which result in hunger, epidemics, population reductions and migrations that sharply affected the development and stability of prehispanic societies.

Chapter organization and content

This work is organized in nine separate chapters. Chapter 2 provides the theoretical background of this work. In it, relevant terminology regarding subsistence agriculture and institutional agriculture is discussed as well as the agro-ecological, cultural and environmental factors that affect crop production.

Chapter 3 deals with the environmental setting of the Tepeaca region. It establishes the natural regions found within the territorial boundaries of the 16th century AD Tepeaca *altepetl*.

Chapter 4 provides a general overview of the region's local culture history. Special interest is placed on the social structure, the land tenure arrangements and the tributary systems prevalent in Tepeaca during the 16th century AD and at the onset of the Spanish arrival.

Chapter 5 provides information on agriculture practices as are performed today within the study region. The goal of this discussion is to establish the energetic input of cultivation tasks and to detect the critical factors dictating agricultural success or failure. Details are given on the agricultural activities for the 2009 agricultural cycle.

Chapter 6 presents the ethnographic field data on regional variability in maize production during 2009. The goal is to establish the patterns of crop production with regards to environmental settings and the years' climatic conditions. It also provides information on the crops cultivated, the size and location of the fields, crop productivity, and family structure within the surveyed zone. This chapter also deals with the drought of 2009 and models its strength and severity within the several environmental zones that comprise the Tepeaca region. Information is provided about the distribution of crop yields in relation to the management of individual household labor and climatic timing.

Chapter 7 examines in detail the characteristics of the landless sector of early Colonial Tepeaca known as *terrazgueros*. These individuals were the main supported for political institutions and the chapter examines changes in the social and tributary relationships that occurred between tributary populations and the elite apparatus during the Early Colonial Period.

Chapter 8 is an analysis of maize productivity for Late Postclassic and Early colonial Tepeaca. It is inferred mainly by employing historical records on field sizes allotted to individual families. Information on environmental unpredictability and unstable maize productivity are added to the discussion in order to establish a more 'pragmatic' view of production values for individual households and regional agriculture. The chapter also discusses the impact of environmental variability on prehispanic food production and on the local tributary system. It also infers several ways in which commoner households might have buffered risk and uncertainty.

Finally, Chapter 9 provides an overview and summary conclusions that can be drawn from this work and the directions for future research.

Appendix A provides the field data on maize productivity collected during 2009 in the Tepeaca Region.

Chapter 2
Agriculture And Theory

The main objective of this chapter is to review and discuss the major social and environmental variables involved in agricultural production. It has two goals. First, I discuss the two food production strategies of subsistence agriculture and institutional agriculture. I explore the goals and functions of these two forms of agriculture because they provide the basis for analyzing the prehispanic agrarian systems of ancient Tepeaca. Their production objectives and the way they operated were interconnected and dependent on each other. Secondly, I examine the environmental, agro-ecological and cultural variables involved in food production. Rain variability, plant interactions, pests, drought and soil composition are problems that must be dealt with each year. In addition, patterns of land and plant use, the specific activities performed by peasants during the cropping cycle, and how decisions are made regarding cultivation tasks, have a direct impact on food production. If it were not for these overlapping and often contradictory sets of variables, the levels of agricultural production could be much higher. Empirically, the only constant for peasants was that every year some sectors of the population would produce well and others badly. It was only under exceptionally good or bad years that production outcomes are distributed uniformly, and even then, there are production sectors that still deviate considerably from yearly mean values.

Subsistence agriculture

I define subsistence agriculture as that strategy in which food production is organized at the household level and destined mainly for the reproduction and survival of its members. Usually, this strategy is self-sufficient because most food is grown and consumed within the household, and it is in this realm in which decision making regarding its distribution, storage, sharing, and pooling takes place. It is remarkable how much of the household's security depends on the adequate management of its resources; this is why food production and its management are based on rational decisions and comparisons of costs and benefits from a series of existing alternatives. These decisions are generally taken in an anarchic way, because the most important facet for the household economy is to satisfy the basic and immediate needs of its members (Sahlins 1972: 95).

Crop production among smallholders is generally low and is geared to satisfy the basic minimum food needs of the household. When farmers are faced with the decision of how many resources they should produce for a given year, they often choose those that will give them the greatest gain with the least effort (Sanders *et al.* 1979: 360). For example, Kirkby (1973: 76) reports that in Oaxaca, when there is plentiful rainfall in midsummer auguring good yields, farmers worked fewer parcels of land because it was expected that the families would get by with the produce of a smaller portion of land. Motivation then, for many farming families was not profit making, but providing the necessary food for the household.

However, why is it that peasants produce for minimum needs and do not seek large profits by cultivating more land? After all, there are indicators that it is not uncommon for households to seek ways in which to increase production (Netting 1993). Small farmers have the capacity to generate innovative technologies and strategies to address the problems and opportunities generated by environmental and economic changes (Brookfield 1972, 2001; Stone 1996). The use of biological diversity and the development of technology and plant managerial techniques are key to household adaptability (Fedick 1996; Sanders and Killion 1992). The question then arises, do households actually have the means to produce more than what they commonly do? And, if not, why is it that their level of production is so close to the household's annual consumption needs?

In fact, there are indications that households often desire to substantially increase their level of food production, but that there are many limitations to their doing so. One factor that appears to limit production in nuclear households is reduced labor availability and how household size and composition varies through its life cycle. Russian researcher Aleksandr Chayanov (1966) introduced the concept of the consumer/worker ratio, which is a measure aimed at explaining the allocation of land and labor by peasant families and how it affects total production (Chibnik 1987; Durrenberger 1980). The number of consumers influences the minimum output that the household must produce, while the consumer/worker ratio influences the amount it will produce beyond this point (Chibnik 1984). Households with high consumer/worker ratios need to spend larger amounts of time meeting their consumption needs. The marginal disutility associated with their labor inputs beyond subsistence is higher than households with lower ratios and thus will tend to produce smaller surpluses. More adult producers in relation to children consumers will achieve a markedly better economic level because they can work more land providing better conditions for surplus generation and food storage. On the opposite side, having more consumers (e.g., children, incapacitated adults) than workers, signifies that the producers have to work harder in order to attain the necessities of the household. Hammel (2005) has pointed out that because Chayanov limited his model to peasant nuclear households and developmental cycles of only 26 years, a useful correction to his formula is to analyze the dynamics of the developmental cycle and the micro-politics of domestic groups. It is possible that the inter-generational structure of complex households and extended kin groups can smooth production and consumption across the domestic cycle of the group and the life cycle of its members.

Households that perform subsistence agriculture may possess several dispersed fields within a close radius from the homestead as well as one near their dwellings. Lands away from households normally do not surpass two to three hectares because the farming size is strongly determined by the amount of land that can be farmed by one farmer (Beets 1990: 384; Kirkby 1973: 73; Logan and Sanders 1976; Sanders and Santley 1983) using simple implements and technologies like those prevalent in prehispanic Mesoamerica (Rojas 1984). A dispersed strategy is a means to buffer the risk of total crop failure and the disparity in the size of plots under cultivation may be related in part to differential access to land, internal social and land tenure arrangements, different economic occupations of their owners and also inheritance rules. In addition, the presence of population pressure on land, resources and agricultural systems, could result in a scarcity of land for cultivation. When Dutch economist Esther Boserup (1965) analyzed the change from extensive to intensification cultivation strategies, she clearly showed that changes in the land use factor are sometimes affected by high population densities and population growth. The scarcity of land available for peasants under this scenario leads inevitably to some sort of change in the intensity of land use. In modern times, land fragmentation problems in central Mexico have been an important factor in creating small plots that in many cases did not surpass more than half a hectare in size. What used to be large fields were split up between the descendants of one original couple, a trend that can be correlated with the recent exponential demographic growth of settlements.

An additional factor is that households may seek to produce only the minimum amount of food because they are engaged in other economic activities such as craft production or wage labor. Having smaller fields could be correlated in some way with restricted access to land, but also because agriculture may not be the principal occupation for the household. Peasants that do not depend strongly on the market economy for provisioning will need to work more land in order to generate the necessary supplies to sustain themselves. Involvement in auxiliary production including crafts, livestock raising, services and wage labor permits the household to obtain additional resources and storable items that can mitigate risk in the future (Hirth 2007; Rijal *et al.* 1991; Shimada 2007). Complementary production of this type can also be an efficient strategy buffering households against crop failure and food shortages due to variable and unpredictable environmental fluctuations.

This risk buffering strategy can also work the other way around. Under some conditions, relying on subsistence agriculture can smooth the problems originating in volatile shifting economic conditions. This is especially true for households engaged in wider commercial networks. By growing their own food, households can be assured of some basic caloric intake and thus impede a total breakdown when an economic crisis occurs. In Belarus and Ukraine, the use of subsistence agriculture has worked as the smoothing mechanism for poor peasant families during periods of economic stresses (Yemelyanau 2009). In Romania (Petrovici and Gorton 2005: 210-211) it is common to observe that households engaged in subsistence food production are more rural and removed from urban areas. This is beneficial for continuing basic subsistence practices because it distances them from agricultural market webs. In Russia, the unstable macroeconomic conditions have generated a shift from large-scale agriculture to semi-subsistence farming (Caskie 2000). Not surprisingly, households will avoid extended periods of dependence on one source of income. They will diversify as much as possible in order to level total productive output. Diversification is an efficient way to manage risk and anticipate adverse conditions (Halstead and O'Shea 1989). In many countries (e.g., Rennie 1991), especially those undergoing rapid modernization such as China (Prändl-Zika 2008), peasants with subsistence agriculture are often pulled into a cash crop economy and practices such as craft production and cattle raising. Despite these changes, many households continue their traditional agricultural strategies in order to produce food for auto-consumption.

In modern monetary economies, households engaged in subsistence agriculture are more common in the rural areas of less developed countries (Thomson 1986) and among non-urban households. Their removal from major economic centers has not prevented them from having a prominent role in cash crop production. This is especially true for communities centered on an exchange based economy. It is now common to observe cash crop agriculture in households including the cultivation of coffee, rubber, medicinal, ornamental plants, spices, as well as weaving textiles. Though cash crops have become an important source of income for communities, societies that rely strongly on subsistence agriculture rarely depend entirely on them as the main source of food and resource acquisition. Nonetheless, they provide additional monetary income and there is no reason to doubt that in antiquity exchangeable products, just like today's cash crops, were also important among prehispanic Mesoamerican societies.

Households clearly can implement a variety of improvements by modifying their natural landscape or through social or economic changes (Brookfield 1972). Most agricultural improvements at the household level are made at a low level of complexity and are highly localized because they deal with local environment adaptations (Wilk 1991). To presuppose that changes are only made possible by a top-down management definitively obscures any analysis of social development and labels peasant households as unskilled and incapable of generating technological advancements which is incorrect.

Another aspect that can lower the level of production, and which is a main object of study in this work, is climatic variability. Because a high percentage of agricultural practices that take place among smallholders depend on rainfed systems, peasants are fully aware that they are limited in their ability to produce large surpluses or profits. Climatic variability negatively affects crop stands and reduces their productive base. Under erratic climatic regimes, energy inputs may be regained at harvest time, and if conditions are optimal, a small surplus may be provided. However, in the majority of cases, surplus over the subsistence needs is relatively low. Data on average maize production under rainfall conditions show that output is highly variable. I have collected a series of reported yields from areas with different environmental conditions and climatic regimes. There are many technical reports presented by agronomists regarding productivity for central Mexican native landraces (e.g., Aceves et al. 2002; Gil et al. 2004; González et al. 2007; Hortelano et al. 2008; Lane et al. 1997; Luna and Gaytán 2001). Unfortunately, they pertain to experimental settings, most under irrigation and with pest controls, and usually report expected productive levels which are extremely high and do not correspond to what fields actually produced under normal cultivation. I have only considered registered final maize productions and not expected ones. Data shows that traditional and developed maize varieties under rainfall conditions yield from as little as 200kg/ha to more than 4000kg/ha. Average yields are of 1232kg/ha (Figure 5).

Location	Reported by:	Maize yields (kg/ha)
Sibak'teel, México	Nigh (1976)*	200
Komchen, México	Shuman (1974)*	250
Lago Petén, Guatemala	Cowgill (1962)*	257
Venda, Transvaal	Piesse et al. (2000)	273
Tlacolula, México	Kirkby (1973)*	380
Lake Petén Itzá, Guatemala	Reina (1967)*	425
Chamula, México	Fernández and Wasserstrom (1977)*	437
Guanajuato, México	Granados et al. (2004)*	500
Chichén Itzá, México	Morley and Brainard (1968)*	524
Chitowa, Rhodesia	Cleave (1974)	587
San Martin, México	Nichols (1987)*	600
Sta. Maria Coatlan, México	Nichols (1987)*	600
Lago Petén, Guatemala	Cowgill (1962)*	623
Mixteca, México	Gaspar (2010)*	650
Quintana Roo, México	Villa Rojas (1945)*	652
Madura, Indonesia	Hoque (1984)	655
British India, India	Blyn (1966)	681
Gaborone, Botswana	Maro (1996)	685
Greater Bengal, India	Blyn (1966)	690
Etla Valley, México	Kirkby (1973)*	700
San Miguel Tocuila, México	Nichols (1987)*	700
Ixil, Guatemala	Monteforte (1959)*	716
Piste, México	Steggerda (1941)*	750
United Provinces, India	Blyn (1966)	796
West Africa	Kowal y Kassam (1978)	800
Succotz, Belize	Arnason et al. (1982)*	800
Shaba, Zaire	ne Nsaku and Ames (1982)	835
Chichipate, Guatemala	Carter (1969)*	847
Salitrón, Copán, Honduras	Shumman (1983)*	871
Lago Petén Itzá, Guatemala	Cowgill (1962)*	878
Greater Punjab, India	Blyn (1966)	904
San Miguel Tlaixpan, México	Báez et al. (1997)*	920
Chiapas, México	Bellon (1991)*	950
Sesemil, Copán, Honduras	Shumman (1983)*	974
Central Oaxaca, México	Kirkby (1973)*	980
KaNgwane, Transvaal	Piesse et al. (2000)	986
Chunchucmil, México	Beach (1998)*	1000
Izucar de Matamoros, México	INIFAP (1997)*	1000

Location	Reported by:	Maize yields (kg/ha)
Malawi	Ng'ong'ola et al. (1997)	1000
San Nicolas de las Ranchos, México	Lopez (2000)*	1000
Xocen, México	Teran and Rasmussen (1994)*	1000
San Salvador Atenco, México	Nichols (1987)*	1000
Tlokweng, Botswana	Maro (1966)	1020
Huehuetenango, Guatemala	Stedleman (1940)*	1037
San Cristobal Colhuacan, México	Nichols (1987)*	1050
Sta. Maria Maquixco el Alto, México	Nichols (1987)*	1050
Los Achiotes, Copán, Honduras	Shumman (1983)*	1074
La Venta, México	Drucker and Heizer (1960)*	1100
Chan Kom, México	Redfield and Villa Rojas (1935)*	1144
Tabasco, Mexico	Gliessman (2000)*	1150
El Cajon, Honduras	Loker (1989)*	1165
San Pablo Ixquitlan, México	Sanders (1957)*	1200
U.S. Native American groups	Schroeder (1999)*	1245
Quintana Roo, México	Villa Rojas (1945)*	1270
Buena Vista Copán, Honduras	Shumman (1983)*	1297
Chile (1944-1945)	De Vries (1952)	1320
San Miguel Tlaixpan, México	Nichols (1987)*	1320
El Cajon, Honduras	Loker (1989)*	1333
San Pedro Chiautzingo, México	Nichols (1987)*	1350
Chile (1955)	Crosson (1970)	1380
Aguacate, Belize	Wilk (1991)*	1393
Lebowa, Transvaal	Piesse et al. (2000)	1399
Nuevo San Juan Chamula, México	Preciado (1976)*	1400
Highlands Chiapas, México	Nigh (1976)*	1500
Midwestern U.S. Experiment 3	Munson-Scullin and Scullin (2005)	1510
Toledo, Belize	Wilk (1982)*	1515
Tierras Altas, Costa Rica	Barlett (1975)*	1533
Zaachila, México	Kirkby (1973)*	1540
Uaxactún, Guatemala	Urrutia (1967)*	1597
San Vicente Chicoloapan, México	Nichols (1987)*	1600
Atlatongo and Calvario Acolman, Mexico	Charlton (1970)*	1613
Zowa, Rhodesia	Cleave (1974)	1716
San Digueto, México	Nichols (1987)*	1725
Indian Church, Belize	Arnason et al. (1982)*	1750
Midwestern U.S. Experiment 2	Munson-Scullin and Scullin (2005)	1824
El Cajón, Honduras	Loker (1989)*	1833
Coatepec, México	Nichols (1987)*	1845
San Martin de las Pirámides, México	Sanders (1957)*	1875
U.S. 1909-1940	Pimentel and Dazhong (1990)	1880
Mount Darwin, Rhodesia	Cleave (1974)	1916
Jerusalem, México	Preciado (1976)*	2200
Midwestern U.S. Experiment 1	Munson-Scullin and Scullin (2005)	2390
San Gregorio Amanalco, México	Nichols (1987)*	2390
Papalotla, México	Nichols (1987)*	2660
Lacanjá Chan Sayab, México	Nations and Nigh (1980)*	2800
San Lorenzo Tenochtitlan, México (1991)	Lane et al. (1997)*	3675
San Lorenzo Tenochtitlan, México (1992)	Lane et al. (1997)*	3761
Tlaxcala, México	Patrick (1977)*	4018
Average		1232
* Employed traditional landraces.		

FIGURE 5. LIST OF 88 AVERAGE MAIZE YIELDS PER HECTARE IN RAINFALL AGRICULTURE REPORTED WORLDWIDE.

Putting things in perspective, according to Sanders and colleagues (1979: 372-373, Table 1) a Mesoamerican household composed of seven people would have needed around 1000kg of maize for their annual consumption if it comprised 50-80% of their diet (see also Sanders 1976). For the Maya area, Steggerda (1941) estimated 1000kg of maize for family of five with a dependence of 80%, while Reed (1998) used a dependence on maize of 70%. A comparison of this figure with the aforementioned average yields indicates a small margin between the production/consumption needs. Native Mexican landraces can produce larger amounts of shelled maize above the necessities of the households, and can reach as much as 4000 to 8000kg/ha under optimal conditions (González et al. 2007: 39-40, cuadro 3). Both agronomists and farmers strive to develop higher production capacities. Empirical evidence shows, however, that this is a wish that in actual fact is rarely achieved due to the severe and often erratic environmental phenomena that cannot be changed.

All these factors affect the productive potential of subsistence agriculture and have important food provisioning issues for households. Although households may desire to produce more than their basic requirements, in the majority of cases, they simply cannot do so due to various natural and social constraints on their agricultural systems. It is only when prominent technological advancements are developed and applied to a better water or nutrient supplies that they are able to generate harvests well above their subsistence requirements. Food production often fluctuates around the minimum necessary for the household's survival and, because of this aspect, it is hard to augment productivity. Instead, output will be low in a lean year with higher levels of production in better seasons. Under these circumstances if households do not have maize in storage or other goods that can be exchanged for food, they face starvation or serious episodes of malnutrition for its members. As a result, food storage systems are an integral part of household survival as is the production of nonperishable items that can regularly or intermittently be exchanged for food when the need arises.

The quickest adaptation to uncertainty or changing circumstances is often at the household level. Households can be very responsive to environmental, economic, ideological, and socio-political changes. Households are often impacted by political and economic institutions but they adapt quickly because they need to survive (Smith 1987b: 297). Unfortunately, unstable economic policies, an overabundance of brokers, recurrent rises in the price of fossil fuels and the insertion of agriculture in a globalize food market have deteriorated the status and income of Mexican farmers. In addition, local problems have created a sharp decline in the economic well-being of people in the countryside. The fragmentation of property, urban development, changes in legal 'land use', reforms from crop raising to mining, impoverishment of fields by overexploitation and erosion, and the desires of families for modern products and services (education, cars, radios, stoves) has led to high indices of land abandonment and migration to cities or to foreign countries, mainly the U.S.A. and Canada.

Institutional agriculture

I define institutional agriculture as the food producing strategies geared towards the support of ruling institutions or apparatus. It can be highly dynamic, politically competitive, growth oriented, and unstable. From the political perspective, it provides the basis for the development of ruling parties. Control over food production is a key feature of chiefdom and pre-capitalist state tributary economies because these institutions require the expropriation of staples and labor to finance stratification and social, political, and religious institutions which are run by non-food-producing personnel (Earle 2000). Within this realm, the objective of agriculture is to maximize the income for ruling elites. Thus, institutional economies evolve in proportion to the access and control over the key resources.

Among agrarian chiefdoms and states, the system of land tenure is seen as the foundation of political control (Johnson and Earle 1987: 15). In general, it is attractive to analyze rural households and communities in relation to ruling entities (i.e., a top-down view) (e.g., Almazan 1999; Brumfiel and Earle 1987; D'Altroy and Earle 1985; Janusek and Kolata 2004). This is due perhaps to the nature of archaeological data in which it is easier to infer elite control over populations through the social, political, economic and religious functions that they perform and that are commonly reflected in the archeological record through mortuary practices, architectural remains, portable goods assemblages or large scale agricultural projects.

Institutional agriculture falls within the type of control over production discussed by Johnson and Earle (Johnson and Earle 1987: 270). This is made possible by the development of technologies such as irrigation or carefully managed short-follow lands. Hydraulic management models state that the key to the political economy lies in elite control over the technology of production. Conversely, other models such as assigned production favor control centered on land and labor (Hirth 1996: 211-213). Both of these two types of control parallel one another because for agricultural production the technological advancements are controlled and used by political institutions.

Political control in Postclassic Mesoamerica focused on control over people's labor. It was manipulated through kinship networks, ritual activity and control of resources or the material means to produce them (Hirth 1996: 205). Agricultural intensification and labor management are seen as central and necessary issues in models of the political economy (e.g., Calnek 1992; Kowalewski and Drennan 1989; Parsons 1991; Sanders and Nichols 1988; Smith 1987a, 1994; Smith and Berdan 1992). It is said that in early civilizations, probably 70 to 90 percent of labor input was devoted to agriculture tasks. This tied households to agriculture and implies that strategies to produce surplus had to come from political entrepreneurs (Trigger 2003: 313-314).

Intensification strategies can be initiated at either the institutional or the household level. Institutional agriculture involves only that production that is controlled and destined for extra-household use and consumption, and as such, is managed by political institutions or ruling groups. Agricultural intensification strategies can be carried out both at this level and at the level of individual household. Developing technological advancements in a particular system is not the most important dimension of institutional agricultural. It is possible that many technological innovations or intensive systems have been developed at the household level but that at a certain point in time were taken over by state apparatus. Such a case appears to have occurred in the development of intensive raised field systems of Lake Titicaca and associated with the consolidation and decline of the Tiwanaku state (Janusek and Kolata 2004).

Subsistence and Institutional agriculture are two different but related types of agriculture. Although they are different, peasant farmers were engaged in both. Peasants were linked to political institutions either by coercion for the payment of tribute or through agreements as renters or attached specialists. Food production is embedded in multiple social domains and through time. Production can shift as a result of changing relations, economic demands, and socio-political circumstances. Yet, even under these ever-changing conditions, subsistence agriculture may not be part of the political realm and will not be interfered with because to do so could jeopardize the production needed to support the population. Political institutions, therefore, rarely interfere with subsistence agriculture. The exception to the rule may be found when conflict arises and food is demanded or taken directly from peasant households. Otherwise, households will determine what to produce and how to use it and will do so without political intervention.

Environmental and cultural variables that affect agricultural production

Agriculture represents one of the most intricate interactions of humans with their environment. This combination is due to the cultural and environmental variables involved in agricultural chores (Figure 6). In this section, I describe the most important variables that shape and restrain agriculture production in both the subsistence and institutional agriculture spheres.

Water availability

Plants require specific thresholds of water to develop correctly and be able to reproduce. How water availability fluctuates ultimately results in variations in grain production and food for humans. Water stress and shortage has been related to problems in plant growth, reproduction, leaf growth area, grain filling, and barrenness (Schulze et al. 2005).

KEY VARIABLES INVOLVED IN A SUCCESSFUL HARVEST	
Cultivation techniques and technologies	Fertilization methods, farming strategies, planting densities, working tasks
Environmental	Altitude, climate, evo-transpiration, soil texture and depth, solar radiation, temperature, water availability
Plagues	Birds, diseases, rodents, scavengers
Plant biology	Allelopathy, competition, facilitation, landrace variety, pollination
Post harvest	Storage facilities, pests, fungi, theft
Other	Cultural and religious practices, plant surviving rate after sowing

FIGURE 6. VARIABLES INVOLVED IN THE AGRICULTURAL CYCLE.

With regards to maize, various factors determine its productivity. In rainfall agriculture, the main restriction is the consistency or inconsistency of water availability (Gimenez et al. 1997: 119). The plant's high sensitivity to water stress, in contrast to other cereals like sorghum or millet (Kowal and Kassam 1978: 255), is the crucial element for maize production and for the development of its different components, especially during critical moments of growth (Adams et al. 1999; Kirkby 1973). Maize is well adapted to rainfed conditions like those prevalent in the Mexican highlands because it employs low levels of moisture given it's elevated transpiration efficiency (Gliessman 2000: 127). Water deficiency results in lower productivity in maize because it retards leaf development that in turn diminishes absorption of solar radiation (Gimenez et al. 1997: 128). In maize, water deficiency is also associated with poor ear formation and development, and under good moisture and nutrient conditions with the proliferation of multi-eared plants (prolificacy) (Svečnjak et al. 2006).

Soils and sediments

Soils and sediments are important in archaeology (Dincauze 2000: 259-260). Pedology, the study of soils in their natural environment, and ethnopedology, the study of indigenous knowledge regarding soil classification and use, have expanded our understanding of prehistoric land use (Wilshusen and Stone 1990). Studying soils helps to understand the taphonomic process to which archaeological materials are exposed, especially perishable materials. Formation processes provide valuable information for reconstructing past environmental change (Schiffer 1996).

In agronomy, soils are simply the surface materials that support plant growth. The quality of soils is relative to the amount of nutrients available for plant development and its texture and overall workability (Instituto Nacional de Investigaciones Forestales 1997; Norman 1979). Soil chemical composition including ph levels and nutrients are used to determine the best crop to be cultivated. The composition and depth of the soil also influence the degree of moisture retention. In particular, the capacity of a soil to absorb and retain water is especially important. Crucial to moisture retention are the hydrological effects of water repellency (Jordán et al. 2009), an aspect not fully studied in central Mexican volcanic soils. Water repellency refers to the ability of soils to resist absorbing water for periods of minutes to hours resulting in reduced infiltration rates, enhanced runoff water and increased soil erosion. Water repellency can be caused by hydrophobic substances released by plants, fungal activity, mineralization, humidification rates, and wildfires; preferential flow paths can lead to variation in water content within soil profiles causing uneven water distribution to the crop root zone which will affect seed germination and plant growth (Willis and Horne 1992).

Sedimentary deposits can be formed by weathering of the bedrock, the transportation and deposition of particles, and post-depositional alteration. Soils, on the other hand, are defined as 'the portion of earth-surface material that supports plant life and is altered by continuous chemical and biotic activity and weathering' (Rapp Jr. and Hill 1998: 18, 30). Sediments are generally biologically dead while soils are active and alive.

Soil depth is crucial for agriculture. Soil degradation is a problem for modern agriculturalist because the root systems are not able to penetrate to a considerable depth that enables them to obtain the necessary conditions for their growth. Shallow soils do not retain moisture and thus plants will not thrive. Deeper soils may retain more humidity and can be less prone to evaporation. However, this will obviously depend on the soil properties and composition. Clayish soils have higher levels of water repellency due to their low permeability and therefore, promote surface water to run off faster. Under extreme conditions, clayish soils will not retain sufficient water for plants. This is the case with tepetate deposits of central Mexico, which is an extremely compacted clayish duripan. Clayish soils will, in some cases, prevent the practice of agriculture at an optimal level. For example, in Chunchucmil (Dahlin et al. 2005) there is sufficient annual precipitation for agriculture (600-900mm) but because of soil chemistry and the fact that soils barely retain water, maize productivity is as low as 100kg/ha.

Sandy soils can be very permeable and good for growing root crops such as potatoes. Nevertheless, they are often too permeable for growing cereals such as maize. Alluvial soils are best for maize because they have good moisture absorption and retention, are easily worked, and require less energy for cultivating than other soil types. Therefore, water retention is equally important for plant growth as water precipitation.

In many places of central Mexico, deforestation for agriculture has led to accelerated soil erosion and alterations to the hydrological balance. Soil degradation and erosion pose a problem to agriculturalists. Strong storms characterized by continuous rainfall usually result in some degree of land degradation. This is especially true for systems under short fallow periods that include a high degree of deforestation, overexploitation, different stages of recovery, and low levels of vegetation recuperation like those associated with slash and burn systems (e.g., Conklin 1954; Freeman 1955; Kass and Somarriba 1999).

In some areas humans have adapted to local settings and have created new spaces for cropping by constructing complex terracing systems. Examples include the Andes (e.g., Zimmerer 1991, 1993), the Maya area (e.g., Murtha Jr. 2002), Madagascar (e.g., Ruthenberg 1971), the Mixteca, Oaxaca (e.g., Feinman and Nicholas 2004), the Philipines (e.g., Dove 1983), Central México (e.g., Fargher et al. 2010b; López 2000), and El Salvador (e.g., Zier 1992). Also, planting rows of trees or maguey plants along the perimeter of fields has been used to construct terraces among Mexican farmers (e.g., Evans 1990; Patrick 1977; Wilken 1987). Yet, soil degradation due to continuous cultivation will inevitably degrade nutrient availability and result in poor yields. Humans have responded to such situations by using different systems of fallow or by artificially fertilizing fields. Fertilization techniques involve a range of decomposing elements that include plants for mulching, animal manure, human night soil, poultry litter, and green waste-compost (e.g., Altieri 1987; Dopico et al. 2009; Kass and Somarriba 1999; Mwalukisa 2008). Night soil (*tlalauiac*), rotten wood (*tlazotlalli*), and forest humus (*quauhtlalli*) were some of the natural fertilizers employed by farmers in Mesoamerica in the Late Postclassic and Early Colonial periods (Sahagún 1963: Book 10). This last technique may be similar to chitemene swidden in Northern Rhodesia (now Zambia) (Kay 1964) in which wood is transported and burned at the field.

Climate variability

Climate variability is a crucial environmental factor involved in crop failures and it relates to the unpredictable nature of the environment. It occurs in forms such as differential precipitation, excessive low or high temperatures, hail, and early or delayed rains. During the Late Postclassic period, moisture deficiency was a major concern for indigenous populations when years of prolonged drought occurred as happened for four consecutive years in Chalco between AD 1332-1335 (Chimalpáhin 1998: 153v.-154v.).[ii]

Rainfall variability is an extremely severe problem for farmers because it produces irregular water availability for plants. It is reflected in differential rain patterning over

---

[ii] "12 Ácatl, 1335. Éste fue el cuarto año en que no llovió sobre los chalcas; y quizá no llovió [en la región], sino sobre las milpas que los tlacochcalcas tenían sembradas entre la [otra] gente. Durante esos cuatro años hubo hambruna, y por eso se cobró [mayor] temor al diablo Tezcatlipoca"

time (within and between years) and space (its distribution within a certain area). This phenomenon is characteristic of tropical climates (Gimenez *et al.* 1997), like those found in highland Mesoamerica. Most precipitation occurs during the summer and at irregular intervals. Storms are copious and highly localized implying different patterns of precipitation at nearby locations (Eakin 2000; Jackson 1978).

The strength and timing of rains have an impact on both soil erosion and water retention. A fine steady rainfall associated with tropical storms allows for a better accumulation of moisture and saturation of the soil. Strong storms are also good, but can cause more damage than benefit by producing gully erosion and plant damage from strong air wind.

Local orography can also play a role in the distribution of precipitation. Orographic uplifting and rain shadows can create differential rain patterns between close communities (McGregor and Nieuwolt 1998: 200-201). Kirch (1994) has reported drastically different rainfall patterns in Polynesia creating different forms of agriculture over very short distances due to the presence of mountain ranges. Geography therefore can influence not only the technologies and techniques employed, but also individual plant management strategies and decision-making.

Rainfall variability also produces problems in correctly identifying the 'timing' or start of the rainy season and the right time for harvesting. All other things being equal, crops sowed at the optimal start of the rains will result in better plant performance and greater yields (Norman 1979). Planting before or after this moment will result in reduced overall production. It is difficult for farmers to determine the right time for clearing land and sowing. Experience may be the key factor in judging the correct 'timing' of agricultural activities (Clark and Haswell 1967). By observing natural phenomena and yearly rainfall patterns, farmers can infer, but hardly predict, the most favorable starting moment. Certain individuals make these judgments better than others and are awarded with a higher status for their special abilities, as are tiemperos in central Mexico (Albores and Johanna 1997; Glockner 1991). At other times, rainfall variability may generate social arrangements that permit access to fields or grazing areas dispersed over wide areas (Thompson and Wilson 1994: 10).

Agro-ecological variables that affect agricultural production

Plant Sowing Densities and Farming Strategies

Strongly articulated to water and nutrient availability is planting/sowing density (Prior and Russell 1975). Studies have shown that fields with high plant densities can have a greater yield, although this depends on having adequate fertilization (Moll and Hanson 1984). Too much density lowers overall yields and leads to high numbers of barren plants due to competition for nutrients. Lower plant densities with good fertilization produce better crop stands. In maize, low plant densities have been correlated with high levels of prolificacy. Also, soils with low moisture retention capacity are best cultivated with relatively low plant densities (de Leon and Coors 2002; de Leon *et al.* 2005).

Monocropping, multicropping, intercropping, and polycropping are different strategies that people employ to meet specific agricultural needs. Monocropping has been linked with modern industrialized agriculture, while polycropping and intercropping are thought of as a subsistence level technique employed to meet household consumption needs (Zimmerer 1993: 19).

Monocropping is probably the simplest form of cultivation, yet it is the most energy demanding under modern industrialized and market oriented economies. It involves cultivating a single crop during the agricultural cycle. Current industrialized cash cropping of coffee, rubber and bananas are some examples. However, monocropping is not restrained to modern or industrial examples. Ancient cultivation of single crops can be found in intensive canal and flooding irrigation of wheat in Mesopotamia and Egypt (Trigger 2003). Modern monocropping systems include the intensive rice terrace systems or sawah in Bali and Borneo (Geertz 1963; Padoch *et al.* 1998), maize farming in areas of sub-Saharan Africa (Kowal and Kassam 1978), grape industrial production in Australia (FAO 2001), and potatoes in Ukraine (Kuchko 1998).

Multicropping has the same characteristics as monocropping in that crops are sowed as stands of a single crop but rotated with other crops in the next agricultural cycle. Generally, multicropping occurs in irrigated zones where two harvests are possible during the year. Cultivars can be rotated between seasons. Alfalfa, peas, or beans may be grown during the winter season and sorghum or maize in the warmer rainy season. In traditional farms, multicropping usually involves the use of legumes interchanged with cereals. Legumes promote nitrogen-producing bacteria on root systems and fix it to the soil replenishing soil nutrients. Cereals are high consumers of nitrogen so cereal-legume rotation is an affordable and 'natural' recycling fertilization system.

Intercropping is a common system in several parts of the world. It involves sowing a series of rows with one crop and another with a different one. Usually only two species are intercropped in a single field, but there are instances that may use three or more. The intercropping technique has two advantages. First, two products can be sown and cultivated in the same field. By this, more products can be raised resulting in a higher Land Equivalent Ratio (LER). The LER refers to the relative amount of an area required under monocropping to produce the harvested yield obtained under intercropping (Vandermeer 1989). In Africa, farmers often plant 10-20 rows of sorghum and next to it 10-20 of a legume during the first season.

During the next cropping cycle, they will shift the order. In some areas of the central Mexican highlands, like Tlaxcala and the Teotihuacan valley, the pattern could be rows of maguey intercalated with wheat, or nopal (*Opuntia* spp.) with maize. In Honduras, Puleston (1982) reports the high potential of breadnut or ramon trees (*Brosimum alicastrum*) that were planted near other crops. In some areas of Puebla and Tlaxcala, farmers may plant up to four diverse crops (maize, beans, cherry and peach trees, as well as the local cherry tree -*Prunus capuli*-) in the same field. The Iban of Sarawak mosaic different kinds of rice within their slash and burn agricultural fields (Freeman 1955). Every rice variety has its own place each year, but surrounding it is a series of other useful plants like fruit and root crops that they cultivate and rotate every year.

It is said that the benefits of intercropping exceed its deficiencies, although it has a higher energetic input/output ratio in comparison to monocropping (Gliessman 2000). In many cases, monocropping is a result of modern economic agricultural orientations. Industrialized monocrops are regarded as more productive, but this may be a fallacy since they use technologies that consume cheap fossil fuels that are very costly to produce and obtain (Mabry and Cleveland 1996).

Intercropping is a useful strategy because it requires less manual fertilization than other techniques. The association of legumes and cereals helps add nitrogen to the soil that the latter require. Yet, maintaining a system such as the Mesoamerican milpa, with its high plant caring requirements (e.g., untangling plants, weeding, and pruning) can be a very demanding job. If chores are not done in time, milpa cultivation may pose problems for agriculturalists and a detriment in overall yields.

Polycropping is similar to intercropping. It involves planting several species intermixed on a single field. Plants are sowed in the same space. For example, climbing beans are put together with maize in the same hole. This way, the bean plants use the maize stalks to crawl and reach sunlight. The milpa system can employ several species (maize, beans, squash, tomato, or peppers) sowed in the same space. In Mesoamerica the tendency for 'traditional farmers' has been the use of polycropping and multicropping techniques (Altieri 1990; Vandermeer 1989) such as in contemporaneous house lots of the Maya area and in los Tuxtlas, Veracruz (Killion 1990).

As with intercropping, some of the benefits of polycropping are facilitation and pest control. Facilitation refers to the benefits that one species such as nitrogen fixing legumes provide for maize. Certain species like squashes around tomato crops act as a barrier and defense for insect pests by discouraging them from penetrating deep into the field. Other plants, such as squashes and eucalyptus, have toxins in their leaf systems which, together with shading, act as a means of controlling the emergence of competing weed plants.

The disadvantages of polycropping include competition for soil nutrients, sun light, and water. Tall fruit trees close to cereals prevent them from absorbing solar energy that can result in poor growth. Cultivating squash alongside maize and beans might not be a good idea if squash vines are allowed to climb, tangle and finally break maize stalks reducing overall yields. Planting two cereals for many consecutively years without appropriate rotation will inevitably exhaust soil nutrients. Maize plants may grow under these circumstances, but very little energy will be available for reproductive functions involving ear development and kernel filling (endosperm). Competition can occur between plants of the same species. The auto-toxicity in alfalfa is a case in point where extremely high levels of toxins inhibit the growth of new plants (Lamb *et al.* 2003).

Cultural factors that affect agricultural production

Subsistence food production depends largely on annual agricultural labor. Land preparation in the central Mexican highlands is known as the barbecho system. Barbecho is usually done using the plow. Plowing land does two things. First, it exposes insects through soil movement and other parasites to bird predators and to extremely low temperatures at night and high temperatures from sunlight during the day. Secondly, the plow loosens soil and incorporates the remains of plants from the previous season as manure. It also ventilates the soil generating good chemical reactions and promoting bacterial activity. The barbecho also helps plant root systems to penetrate the soil in order to obtain optimal moisture. Additionally, weeds can be removed efficiently reducing competition between crops and other plants.

Seed selection is another crucial factor for crop cultivation. Maize agriculture in Mesoamerica, in contrast to Old World agriculture, requires a lower seed input to achieve a normal harvest. About 20 to 25kg of maize is required to sow one hectare with a density of 30,000 plants. Under subsistence agriculture, such densities will generate around 1500kg of shelled maize per hectare. According ethnographic data collected by Angel Palerm (1990: 45, 49, 53), maize can generate between 1/100 to 1/200 seed input/output ratios. Thus, the seed required for the next agricultural cycle is only 1/50 to 1/75 of total production, or even lower. This is a much lower ratio than wheat or sorghum, which require about 1/7 to 1/8 of the previous harvest production.

During the Late Postclassic and Early Colonial periods maize seeds were selected from individual ears. The Florentine Codex (Sahagún 1963: 283) mentions that:

> The best seed is selected. The perfect, the glossy maize is carefully chosen. The spoiled, the rotten, the shrunken falls away; the very best is chosen. It is shelled, placed in water. Two days, three days it swells in the water. It is planted in worked soil or in similar places.

The characteristics of each maize type and that of the different plant components including leafs and kernels are very specific. Characterizations such as big, thick, long, large, slender, spongy, hard, small, hardened, transparent and ruddy are used to describe kernels in prehispanic central Mexico (Sahagún 1963: 279-281). Unfortunately, these texts only mention selection practices on kernels and cobs. Yet, maize can be selected on the basis of the plant overall attributes. Canes for sugar (*pinolli*), thicker or larger husks for tamale, and plants more or less tolerable to salinity or waterlogged conditions are examples of different attributes targeted for selection. For example, (Molina 2008 [1571]) tells that maize was categorized on the basis of the number of days it took to grow and produce; *xiuhtoctlaulli*, for instance, a 50-60 day maize variety.

Knowledge of the different native varieties almost certainly developed over centuries through conscious genetic selection. The Florentine Codex (Sahagún 1963: 282) classifies maize varieties according to the type of environment where they were sown, clearly distinguishing between maize for dry lands and that for use in chinampa or wetland agriculture. In modern times farmers from the Puuc region of the Maya lowlands consider local topography and drainage characteristics for selecting between 30 different native maize varieties for planting (Dunning 1996).

As mentioned before, a crucial element in agriculture is the timing of the rainy season. Like today, Late Postclassic populations probably tried to determine the best time for initial sowing tasks and other activities. Knowledge off environmental phenomena was crucial to predict the start of the rainy season. Sahagún (Sahagún 1963: Book 7, Ch. VII) explained that 'When [clouds] billowed and formed thunderheads, and settled and hung about the mountain tops, it was said: the Tlalocs are already coming. Now it will rain'.

Fertilization methods have become the 'Aquiles heal' of modern agriculture. The use of fertilization today often depends on highly variable fertilizer prices which fluctuate with macroeconomic pricing of petroleum products. Like other countries where subsistence agriculture is practiced (Yiriode *et al.* 2006), fertilizer prices in Mexico are both unstable and often too expensive for subsistence agriculture. As a result, farmers tend to avoid buying fertilizer, and when they do, they buy them only in small amounts.

Storage is a crucial buffering strategy against recurrent environmental variability and food shortages. Harvesting and calculating the available supply of food is part of the food production cycle. It all comes down to what a household can retain and secure for future needs. Storage and food processing, therefore, are key factors in determining the 'real' amount of food available in a given year for the household. Much of the harvest may be lost to pest infestations, spoilage, waste or fungus that occur post-harvest (Schroeder 1999; Smyth 1989). In fact, yield should be measured in terms of what is actually consumed and not what was collected at harvest because inadequate storage and food processing can lead to important food losses. Pests may destroy plant stands in the field, but other plagues and fungi can do the same in poor storage facilities.

Agricultural populations have tried to avoid pre and post-harvest loses by developing several ways to control pests or to preserve foodstuffs after they are collected (Sinclair and Gardner 1998). Preservation technologies include the use of bins, jars, granaries, silos, underground pits, and bamboo containers, as well as treatment of products with preservatives, pre-cooking and oven heat. In Mesoamerica, common forms of storage were underground pits, bins made of pine wood called cincalli which were high in odorous resin to repeal insects, granaries called cuexcomates, and ceramic pots (Hernandez Xolocotzi 1965; Seele and Tyrakowski 1985); likewise, pulverized saline compounds derived from sediments, such as *tenextli* and *tequizquitlalli* among Nahuatl groups, are commonly put into storage facilities due to their ability to repel small insects.

The land use factor is related to both agricultural productivity and the level of intensification. It refers to the relative intensity of how lands are used. Several land use factor classifications have been proposed and applied in Mesoamerican research, including the models of Esther Boserup (1965), Angel Palerm (1967), and Eric Wolf (1966) (Figure 7). During the Late Postclassic, the Tepeaca region had relatively large and dense populations. As many as 250,000 persons may have lived in the surrounding Tepeaca region giving a density of 480 persons/km^2 (Cook 1996: 113). The Tepeaca settlement survey (Anderson 2009: 86-88) suggests that most settlements during this period were rural communities. The historical data show that the land allotment to individual households consisted of 5 to 7 parcels of land (López Corral and Hirth 2012; Martínez 1984a), each one in the order of 0.17ha in size. This indicates a certain degree of land fragmentation. Because of both high population densities and land fragmentation, it is probable that a proportion of the landscape was devoted to permanent cultivation or annual cultivation regimes. It is also possible that the Late Postclassic situation may have looked fairly similar to today's low production capacity and declining marginal productivity (Abrams 1995). This is particularly notable when poor or no fertilization methods are employed. Continuous cropping with two cultivations cycles (and two harvests) could have been practiced in plots set near the Barranca del Aguila where irrigation could be practiced as we see today in the area surrounding the town of Xochiltenango.

For archaeologists, agronomic features present a socio-political view of a society, and provide a good way to measure the productive capacities of an agricultural system. The problem arises when it becomes clear that a wide range of technologies and cultivation techniques

Boserup (1965)	Palerm (1967) (1972)	Wolf (1966)
1. Forest-follow cultivation: (1-2 years of use: 20 years of follow)	1. *Roza* (2-3 years of use: 10-12 or more of follow)	1. Swidden long-term following systems
2. Bush-follow cultivation (1-2 years of use: 6-8 years of follow)	2. *Barbecho* (2-3 years of use: 2-3 years or more of follow)	2. Sectorial following systems (2-3 years of use: 3-4 years of follow)
3. Short-follow or grass-follow cultivation (1-2 years of use: 1-2 years of follow)	3. *Secano intensivo* (annual)	3. Short-term following systems (1-2 years of use: 1 year of follow)
4. Annual cropping (1 harvest a year)	4. *Humedad y riego*	4. Permanent cultivation hydraulic systems
5. Continuous cropping (2 or more harvests a year)		5. Permanent cultivation of favored plots, infield-outfield systems

FIGURE 7. THREE DIFFERENT PROPOSITIONS FOR LAND USE FACTORS.

may be employed for either subsistence or political strategies. Technological features, therefore, need to be analyzed within their broader socio-political contexts and archaeological data. Encountering irrigation features does not imply the presence of state control over production; it is just as likely that independent or cooperating households with intensification goals developed hydraulic features.

Mesoamerican agriculture: rainfed dependent and artificially supplied water system

For this work, I make a basic distinction between rainfed agriculture (RA) and artificial water agriculture (AWA). These two forms of agriculture are radically different in terms of their production capacities. This basic division is the starting point for distinguishing low-level production from high yield agricultural systems in the Mesoamerica.

Rainfed agricultural systems depend entirely on rainfall as the main supply of water for plant cultivation. In the tropics, rainfed systems are implemented during the irregular rainy season, which is the most important variable restricting agricultural productivity. Most traditional maize agriculture in Mexico, as in the majority of Middle America countries (Whitmore and Turner II 2001: 22), is of rainfed type. This probably was also true for agricultural systems during Late Postclassic and Early Colonial periods in central Puebla and Tlaxcala (Trautmann 1997). The risky nature of rainfed agriculture has an important impact on the development of social complexity. The irregular water supply to fields creates a mosaic of variable productive capacity among peasant households within a region. This variability has been recognized as an important factor in the history and development of ancient populations. As proposed in Mesopotamia (Stein 2004) and the valley of Oaxaca (Kowalewski and Drennan 1989), areas that have relatively good and constant precipitation regimes can generate longer and more productive settlements compared to areas with low and/or variable precipitation.

On the other hand, artificial water agriculture provides improved soil moisture either by supplying or removing excess water from fields. It is exemplified by canal irrigation systems (found almost everywhere), raised fields (Tlaxcala, Yucatan Peninsula), chinampas (Basin of Mexico), and drained fields (Yucatan Peninsula, Veracruz, Cholula, Lake Titicaca). AWA systems are more reliable, stable, productive, consistent, and energy efficient because water input is efficiently controlled (Mabry and Cleveland 1996: 232-233). Water supply, sowing tasks and the 'timing' of cultivation tasks are controlled better, thus reducing risk due to climatic variability and water availability (Cleave 1974). Where rainfed agriculture struggles, AWA systems thrive where conditions permit.

As simple as it may seem, having lands on either AWA or RA can make a considerable difference. Some researchers like Parsons (1992), Siemens (1983), and Sanders (1957) have argued that AWA systems are as much as five times more productive than RA. However, perhaps the real advantage of AWA is not just its elevated production potential; it has the advantage of the greater stability and predictability it brings to the farmer. This stability augments the production of staples, and, even if it entails more work for the household farmers, they know that by harvest time they will have a much better chance of getting a good crop.

AWA systems have the possibility of producing two crops per year. The advantage is that if crops are lost due to frosts, fires, plagues, or disease, the farmer can recover by re-planting and generating a second harvest. Certainly, the production of the second crop will be different for the summer and winter seasons. Solar radiation in the summer is much higher and C4 type plants such as maize and amaranth thrive better. Likewise, in the winter lower temperatures and fewer sunny days promote a shorter growing season resulting in a decline of production capacity. Nonetheless, the advantages of AWA are unsurpassed in contrast to RA systems which depend on a once a year precipitation regime.

Today, Mexican agriculture operates under two cycles. The first cycle is Spring-Summer (S-S) and the second is Autumn-Winter (A-W). Maize productivity during S-S is generally higher due to the optimal environmental conditions under which crops can be raised, including more sunny days, higher radiation levels, and fewer frost events. Maize can be cropped in A-W, but the productivity is on average around 30% lower. In A-W, sometimes other crops are raised that perform better in cooler irrigated conditions such as wheat and a variety of cash crops composed mainly of vegetables and flowers.

Agricultural food shortage

Worldwide strategies for coping with food shortfalls have been amply discussed in the anthropological literature. Response mechanisms include building social networks and developing exchange, storage and diversification strategies in both the political and subsistence economies. Sharing, pooling, or relying on less desired foods (Minnis 1985) may be options for household members under heavy food stresses or shortages. Starvation can drive farmers to eat wild resources including those considered to be low quality and/or poisonous. Prolonged drought and worsening environmental conditions can promote transformations in ritual activity. An example of this seems to have occurred during a drought episode during Late Classic period in the Maya Area when a fertility cave cult appeared (Moyes *et al.* 2009).

Peasants have developed several ways to cope with water retention problems and to mitigate the effects of seasonal droughts. Small everyday tasks that can bear important responses to reduce unpredictability and encourage higher production do not appear in the archaeological or ethnographic record. Some of these include basic chores like mulching, seedling transplanting, terracing and polycropping. Others include improved decision making, using the accumulated experience regarding natural phenomena (e.g., planting in relation to the timing of rains) or implementing labor intensification strategies at the household level (e.g., personalized rearing of plants, irrigation).

Unfortunately, even these strategies are no match for extreme rainfall variation or prolonged droughts. For instance, the 'canícula' phenomenon or intraestival drought (see Chapter III) that occurs in midsummer over the central highlands produces a regionally anticipated drought that commonly lasts one to four weeks. In some seasons, this phenomenon can extend for a longer time, as it happened during 2009 when the dry period prevailed for more than two months.

Climatic events such as El Niño-Southern Oscillation (ENSO), a phenomenon brought about by semi-periodic changes in the tropical Pacific Ocean atmosphere system (Adams *et al.* 2003), the North American Monsoon System (NAMS) (Douglas *et al.* 1993), and the Intertropical Convergence Zone of the East Pacific Ocean (Magaña *et al.* 2004) can influence the length of the canícula. For Mexico, in particular ENSO can produce additional or prolonged drought or flood events (Conde *et al.* 2006) and thus can have a strong impact on agriculture production.

These climate disruptions can strongly affect crop failure. In Zimbabwe, a country that has adopted maize as one of the main cereals, more than 60% of the variance of maize yields is explained by sea surface temperatures in the eastern equatorial Pacific related to the ENSO. Historically, there are several regional of focalized drought episodes in Mexico that appear at a constant rate (García *et al.* 2003). New tree-ring data obtained from native stand of Douglas-fir (*Pseudtosuga menziesii*) from La Fragua, in the northern State of Puebla shows that harsh drought episodes have taken place in the central Mexican area within relatively short spans of time, some of which are extremely important (Therrell *et al.* 2006). Due to climate variability, it is estimated that as much as 90% of modern effects to agriculture in Mexico come from drought (Appendini and Liverman 1994: 156). Peasants have had to adapt to the unpredictability of drought. Although they are unaware of when drought will strike, they are certain that it will arrive sooner or later.

Understanding the unpredictable nature of seasonal rainfall and its impact on agricultural systems is crucial for reconstructing prehispanic food provisioning patterns at both the household and institutional levels. Food shortages associated with seasonal and inter-annual climatic variability can be divided into two broad types: localized and regional. Localized Agricultural Food Shortages (LAFS) are mainly the result of Rainfed Agriculture systems characterized by uneven precipitation across time and space. This variability shapes seasonal crop production, generates some degree of food shortfall, and is responsible for inter-annual yield disparities within regions. LAFS produce 'normal' variation in annual food production that put households at risk of failure. Either the household produces enough food to sustain the family for the year, or there is some level of shortage resulting in the need to obtain resources by other means such as sharing, redistribution, exchange, or negative reciprocity.

LAFS usually strikes among smallholders practicing subsistence agriculture. Cultivating a few small rainfed fields in a relatively short radius from the homestead is not enough to mitigate the effects of seasonal climatic variability. Production strategies are different for institutional agriculture because food acquisition is more consistent given the large territorial estates over dispersed areas that tributary subjects cultivate for elite. The state often controls more productive AWA lands such as irrigated areas or wetlands which average out overall yields and cope with seasonal shortfalls. It is only under rare situations, such as prolonged droughts or conflict, when food production for the institutional economy level could be at risk.

On the other hand, Regional Agricultural Food Shortages (RAFS) are events that strike large areas and several populations. They can be caused by prolonged droughts or seasonal events such as the dry canícula. RAFS cause widespread crop failure and, under a series of continuous bad years, can result in high levels of population decline and out-migration. Examples of extreme RAFS events were the prolonged three and four year droughts that struck the central Mesoamerica highlands between AD 1332-1335 (Chimalpáhin 1998: 153v.-154v.) and AD 1454 to 1456 (Durán 2006 [1579]: Chapter XXX).

For the Tepeaca region, there are only a few recorded RAFS events of this nature. There are clues about regional droughts recorded in the *Anales de Tecamachalco* for the towns of Tecamachalco and Acatzingo (Solís 1992). Several other historical food shortages in central Mexico associated with drought periods are provided by García and colleagues (2003). Unfortunately, yearly RAFS linked to inter-seasonal climatic variations, like the dry canícula, are hard to detect in the archaeological record. The reason is that these kinds of shifts do not necessarily correlate with total annual precipitation records. As I show in Chapter VI, overall precipitation for any particular year may reach the mean historical average, while the pattern of rains may vary significantly when they occur. Even a tree-ring based analysis, as the one currently developed for the central area of Mexico, may not register these short term rainfall variations.

In this chapter, I have outlined the main characteristics of subsistence and institutional agriculture. In addition, I have showed that agriculture practices embrace a complex set of variables and factors. These aspects need to be clearly outlined in order to understand the production capabilities of ancient agricultural systems. Probably, the most important aspect discussed here is that agricultural productivity can be highly variable and hard to predict. However, it is this variability that makes it of interest to researchers and so important in understanding food provisioning systems and their effect on social development and economic relations among prehispanic populations.

Chapter 3
The Natural Setting

The study region lies within the Neo-Volcanic axis of the central Mexican plateau, a macro-province with a large and complex geological history. Volcanism and tectonics have produced in central Puebla a series of volcanic formations and elevations —the highest found in Mexico. Extended mountain ranges circumscribe alluvial valleys creating different environmental regions. Within the 16th century AD territorial boundaries of the Tepeaca *altepetl*, sections of three regions converge: the Tepeaca valley, the southern portion of the Llanos de San Juan, and the eastern portion of the Puebla-Tlaxcala valley (See Figure 2). The northwest-southeast mountain chain formed by the Acajete plateau and the Tezontepec and Cuesta Blanca ranges separate the Tepeaca valley from the northern Llanos de San Juan.

The Acajete Plateau and an east-west mountain chain known as the Sierra de Amozoc-Tepeaca separate the Puebla-Tlaxcala valley from the Tepeaca valley. In this area, both elevations form a natural passage around 5km wide into the Tepeaca valley plains. Situated outside the study area, the El Tentzo mountain range forms the southern limit of the Tepeaca valley. The eastern section is delimited by the Palmar de Bravo valley located east of Quecholac and Tenango.

The geographical and geo-statistical data used to construct the region's base was obtained from Mexico's *Instituto Nacional de Estadística y Geografía* (INEGI) (National Institute of Statistics and Geography). The information was downloaded from their website (www.inegi.gob.mx) including state and municipal political boundaries, as well as environment, climate, geology, soil, hydrology and infrastructure information. Aerial orthophotos were provided by the GIS and Mesoamerican Labs at the Department of Anthropology at Penn State University and are at scale 1:20,000. Digital elevation (.bil extension) maps and shape (.shp extension) files can also be downloaded from INEGI and can be accessed and modified through ArcGIS software. From these files, topographic maps were drawn using 50 to 100 meter contour lines.

Soil maps were also obtained from INEGI (Figures 8 and 9). Although an important and highly detailed data set on regional soils was developed by the German Puebla-Tlaxcala project in the 1970s (Erffa *et al.* 1977, Werner 2012), it unfortunately only covers part of the regional boundaries used in this work. Therefore, I have opted to employ the broader INEGI dataset that corresponds to the FAO Soil Units classification (http://www.fao.org/nr/land/soils/key-to-the-fao-soil-units-1974/en/) because it covers all the Mexican territory. Relying solely on the German maps would undermine any comparison that could be made with sites or regions outside the Puebla and Tlaxcala states.

All annotations, including the fields and towns surveyed, were plotted onto maps using the same symbolic notation. The data and resources employed for establishing the different environmental characteristics (e.g., soils, humidity retention, evaporation, precipitation) of each area were taken from both INEGI and the *Comisión Nacional del Agua* (CNA) (National Water Agency) at its website (http://smn.cna.gob.mx/smninfo/mapasitio.html). The CNA has climatic records for the years 1971 to 2000. Fortunately, recent data on precipitation and weather conditions for five stations can also be acquired from the *Instituto Nacional de Investigaciones Forestales Agrícolas y Pecuarias* (INIFAP) (National Institute of Forest, Agricultural and livestock Investigations) website http://clima.inifap.gob.mx/redclima/, which, in the majority of cases, has records from 2005 to 2010.

Another excellent source of data is the *Atlas Climático Digital de México* provided by the *Unidad de Informática para las Ciencias Atmosféricas y Ambientales* (Informational Unit for the Atmospheric and Environmental Sciences) of the National Autonomous University of Mexico (UNAM) with records from 1950 to 2005. The site can be accessed at http://www.atmosfera.unam.mx/uniatmos/atlas/. Unfortunately, the data can only be viewed on screen and cannot be downloaded or printed. Although still under development, the *Puebla's Clima y Urbanización en el Valle de Puebla* Project from the *Benemérita Universidad Autónoma de Puebla* also has some useful complementary information at their website http://urban.diau.buap.mx/index.html.

Climate varies micro-regionally, but in general it sticks to a sub-humid C (w1) (w) and C (w2) (w) type. Rainfall presents a unimodal distribution with a single dry and wet season regime (McGregor and Nieuwolt 1998: 202-204). Rains are concentrated mainly during the warmer summer. The south-central area is a transition zone between the cooler-wetter northwest and the drier-warmer south. These conditions are strongly influenced by the Sierra Madre Oriental, and the Sierra Madre del Sur, which together constitute a natural barrier for humid air currents arriving from the east. Scattered showers occur in the cooler winter season and account for only 5% of the year's precipitation. Substantial showers begin in April-May and become consistent by June and July. Summer characteristically has intense and irregular storms, usually during late afternoons-early evening. In July-August, high pressure can develop over the central plateau disrupting

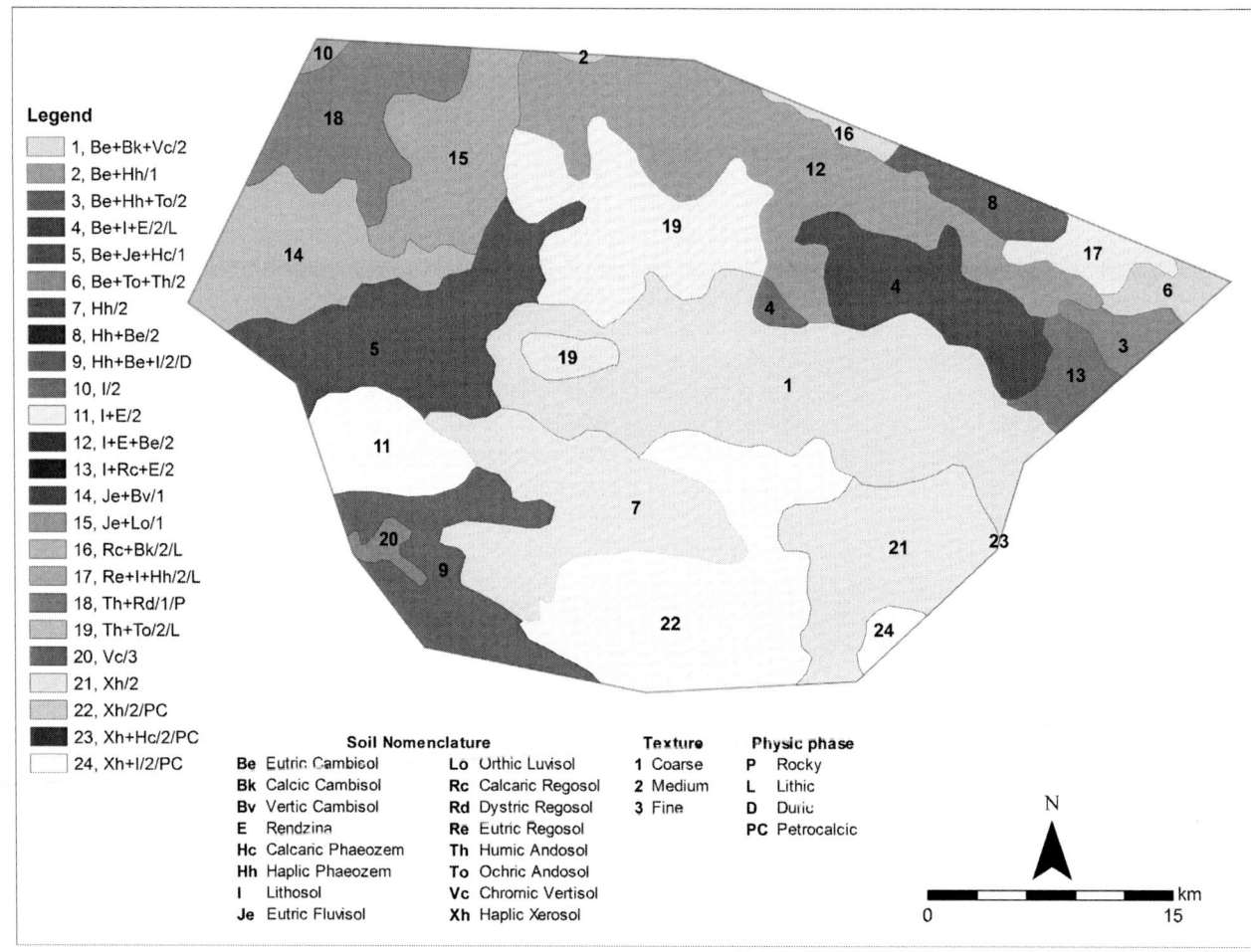

Figure 8. Soil map of the study region.

ANDOSOLS (T)
Other soils having either a mollic or an umbric A horizon possibly overlying a cambic B horizon, or an ochric A horizon and a cambic B horizon; having no other diagnostic horizons (unless buried by 50 cm or more new material); having a depth of 35 cm or more one or both of: (a) a bulk density (at 1/3-bar water retention) of the fine earth (less than 2 mm) fraction of the soil of less than 0.85 gr/cm and the exchange complex dominated by amorphous material; (b) 60 percent or more vitric volcanic ash, cinders, or other vitric pyroclastic material in the silt, sand and gravel fractions.
Humic Andosols (Th)
Other Andosols having an umbric A horizon
Ochric Andosols (To)
Other Andosols having a smeary consistence and/or having a texture which is silt loam or finer on the weighted average for all horizons within 100 cm of the surface.
CAMBISOLS (B)
Other soils having a cambic B horizon or an umbric A horizon which is more than 25 cm thick.
Vertic Cambisols (Bv)
Other Cambisols showing vertic properties.
Calcic Cambisols (Bk)
Other Cambisols showing one or more of the following: a calcic horizon or a gypsic horizon or concentrations of soft powdery lime within 125 cm of the surface when the weighted average textural class is coarse, within 90 cm for medium textures, within 75 cm for fine textures; calcareous at least between 20 and 50 cm from the surface.
Eutric Cambisols (Be)
Other Cambisols
FLUVISOLS (J)

Other soils developed from recent alluvial deposits, having no diagnostic horizons other than (unless buried by 50 cm or more new material) an ochric or an umbric A horizon, an H horizon, or a sulfuric horizon.
Eutric Fluvisols (Je)
Other Fluvisols
LITHOSOLS (I)
Other soils which are limited in depth by continuous coherent and hard rock within 10 cm of the surface.
LUVISOLS (L)
Other soils having an argillic B horizon.
Orthic Luvisols (Lo)
Other Luvisols
PHAEOZEMS (H)
Other soils having a mollic A horizon.
Calcaric Phaeozems (Hc)
Other Phaeozems being calcareous at least between 20 and 50 cm from the surface.
Haplic Phaeozems (Hh)
Other Phaeozems.
REGOSOLS (R)
Other soils having no diagnostic horizons or none other than (unless buried by 50 cm or more new material) an ochric A horizon.
Calcaric Regosols (Rc)
Other Regosols which are calcareous at least between 20 and 50 cm from the surface
Dystric Regosols (Rd)
Other Regosols having a base saturation (by NH4OAc) of less than 50 percent, at least in some part of the soil between 20 and 50 cm from the surface.
Eutric Regosols (Re)
Other Regosols.
RENDZINAS (E)
Other soils having a mollic A horizon which contains or immediately overlies calcareous material with a calcium carbonate equivalent of more than 40 percent (when the A horizon contains a high amount of finely divided calcium carbonate the colour requirements of the mollic A horizon may be waived).
VERTISOLS (V)
Other soils which, after the upper 20 cm are mixed, have 30 percent or more clay in all horizons to at least 50 cm from the surface; at some period in most years have cracks at least 1 cm wide at a depth of 50 cm, unless irrigated, and have one or more of the following characteristics: gilgai microrelief, intersecting slickensides or wedge-shaped or parallelepiped structural aggregates at some depth between 25 and 100 cm from the surface.
Chromic Vertisols (Vc)
Other Vertisols
XEROSOLS (X)
Other soils having a weak ochric A horizon and an aridic moisture regime; lacking permafrost within 200 cm of the surface.
Haplic Xerosols (Xh)
Other Xerosols

FIGURE 9. FAO SOIL UNITS DESCRIPTIONS.

the easterly flow of winds and giving rise to a lull period or intraestival drought known as the dry canícula which normally prevails from for one to four weeks (Magaña et al. 1999). The rainy season decrease in October and return to a pattern of scattered showers in November. Rainfall patterns tend to be localized with certain areas having important accumulations of precipitation on most days.

Annual precipitation increases from east to west and south to north (Figures 10 and 11). According to INEGI (2000) and the *Comisión Nacional del Agua* (CNA), on average, the southern Tepeaca Valley receives 600mm during the summer months while the piedmont of the Malinche volcano gets up to 900mm. In addition, there is a rise in rainfall with greater orographic elevation. The upland piedmont areas above 2500m usually receive more water than the drier and lower alluvial areas of the southern portion of the Tepeaca valley. The eastern portion registers less accumulation than the west due in part to the rain shadow produced by the Citlaltepetl or Pico de Orizaba volcano and the Sierra Madre Oriental. For the entire area, the greatest accumulation of rainfall (1000mm) is registered in areas closest to the La Malinche volcano and the lowest (500mm) in Tecali, Tecamachalco, Quecholac, and Atenco (Figure 10).

Rarely does the rainy season fail, but the beginning of the rains is frequently delayed by a month or more; alternatively it may arrive weeks before expected (Figure 12). Rainfall is highly variable between years and commonly falls below the requirements for good maize cultivation. It is not rare to see common micro-droughts that produce water deficits during plant growth.

The dry winter period is cool with sporadic frosts and occasional rainfall as the result of cold polar air masses that surge southwards, also known as nortes (Liverman and O'Brian 1991: 354). The relative timing of the rains and the occurrence of frosts are major problems for peasants. The incorrect timing of initial sowing combined with early frosts can be fatal for crop stands. Frosts are a severe problem and can occur between October and March with the highest incidence happening in December-January. The number of frosts can be extremely variable, but certain areas are more susceptible than others. In the Amozoc vicinity, the incidence can be of up to 100 frosts per year. In the Tepeaca region, the problem is less serious but still there can be 40-80 frost days. In Acatzingo, and in areas further south at or below 2100m, there may be less than 40 frost days (INEGI 2000). For this reason, production varies greatly between areas and among individual smallholders with adjacent fields.

Hailstorms are another problem for agriculturalists, although not as severe or as frequent as frosts (Fuentes 1972). Hail occurs mainly from March through June and thus can affect plants during their early development or reproductive stages. The chances of a field being hit by hailstorms are normally very low, and weather records show that they occur less than one per day per year on average. However, in 'real life' hailstorms may be highly variable: some areas may be free of hail for several consecutive years while others may suffer their effect on a series of subsequent years or more than once in the same cycle.

The lower alluvial plains and large portions of the piedmont today lack forest vegetation and have been converted to agricultural fields. In some case, the piedmont areas have been affected by erosion (Figure 13). This has turned out to be a serious historical problem in some areas to the point that hills in Tepeaca, Obregón, and Tecamachalco are deprived almost entirely of forest covers and soil accumulation is minimal (Figure 14). Plants that grow in these settings include several types of maguey (*Agave* spp.), *nopal* (*Opuntia* spp.), *biznaga* (*Mammillaria* sp.), *izote* (*Yucca* spp.), *Hechtia roseana*, as well as small shrubs and other wild grasses (Rzedowski 2006: 259, 265).

In the Amozoc, Quecholac and Tenango hills there are still important patches of oak and pine forest. In areas above 2800m, mainly on the la Malinche volcano, forest cover is dense and consists mainly of pine (*Pinus* spp.) along with other cooler climate tolerant species (Klink 1973) (Figure 15). The majority of these forests have been intensively exploited for lumber, fuel, charcoal, or have been converted to grasslands for bovine and equine animals. In the Acajete plateau and the Tezontepec and Cuesta Blanca mountain ranges, there are important areas with several species of oak forests (*Quercus* spp.) and associated communities (Figure 16); Above 3000m, there is an evergreen forest composed of pines (*Pinus* spp.), cypress (*Cupressus lindleyi*) and fir (*Abies religiosa*), especially in gullies or barrancas (Niembro 1986; Rzedowski 2006: 290,308-309,322).

The Tepeaca valley region

The Tepeaca valley is an important natural communication corridor leading to other regions like the Tehuacan valley, the Oaxaca region and the Pacific Ocean. Elevation in the central plains of the valley ranges between 1900 and 2200 meters. Virtually all the central valley floor is under a permanent cultivation regime using both rainfall and irrigation agriculture. In some areas of Acatlán, Tenango, Tlayoatla and Villanueva, agricultural fields have replaced the natural vegetation on the lower piedmont portions of hills. Unfortunately, this has led a rapid erosion of the local soils and has generated sever problems of land degradation.

Climate varies within the region and temperature is clearly influenced by elevation but, in general, it is adequate to support the cultivation of both six month and four maize varieties. Maize is usually cultivated with other products such as beans (*Phaseolus vulgaris*), squash (*Cucurbita* spp.), and *ayocote* (*Phaseolus coccineus*). Sorghum and many types of commercial vegetables are grown in irrigated areas. The south strip, comprising the towns of Tecamachalco, Tecali, Hueyotlipan and Cuauhtinchan, is lower, warmer, and drier. Mean annual temperatures

The Natural Setting

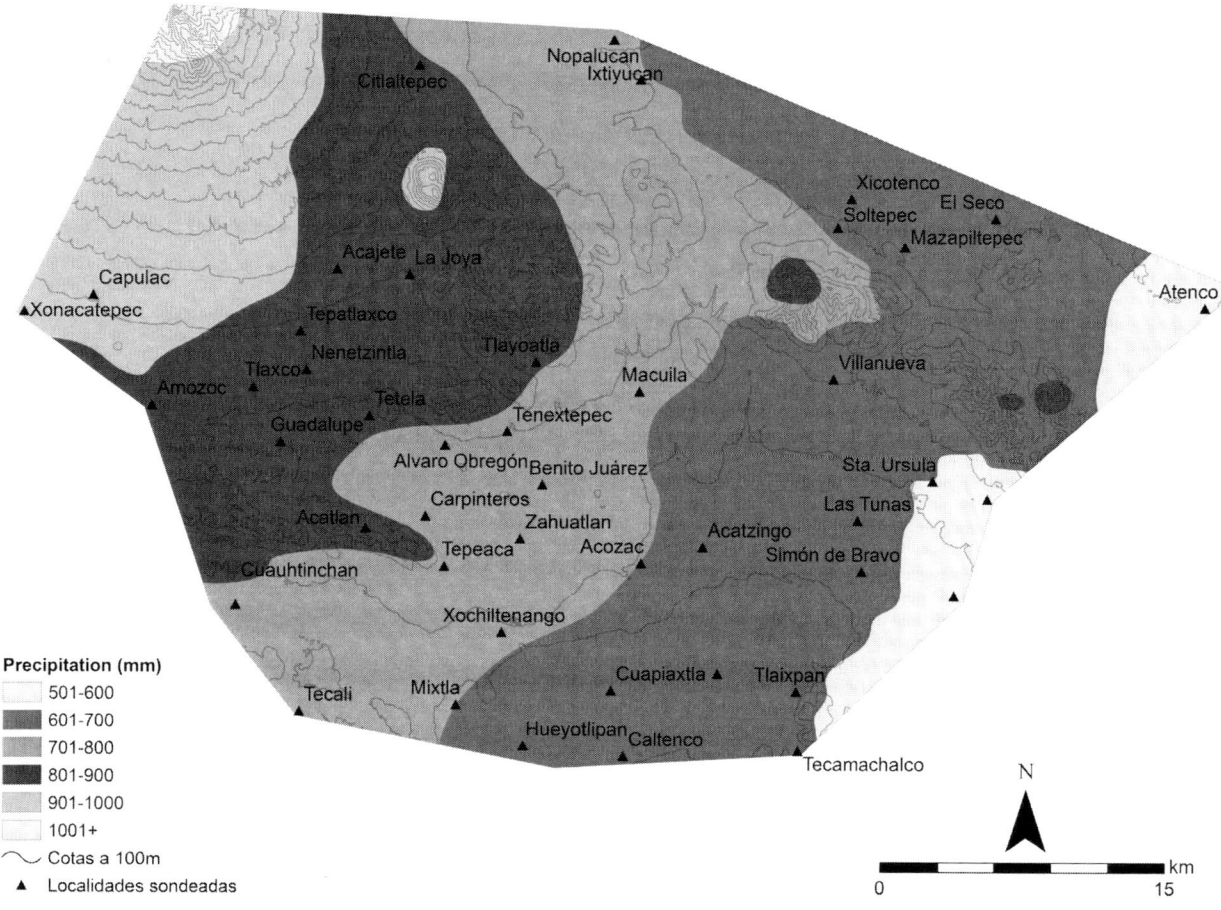

Figure 10. Spatial distribution of rainfall totals within the study region.

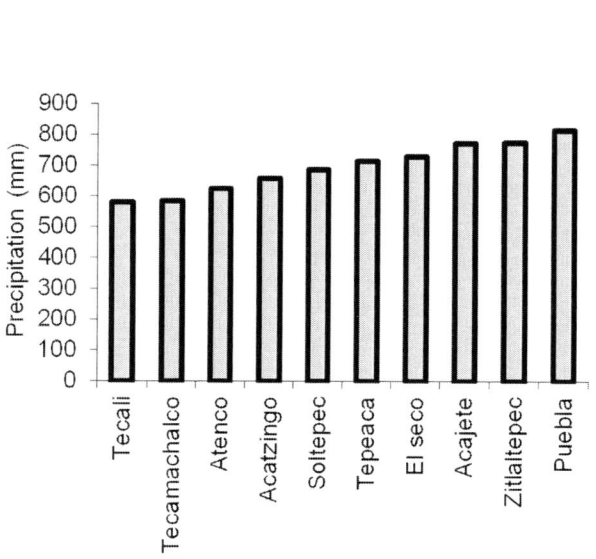

Figure 11. Mean precipitation recorded from 10 different climatology stations from the studied region.

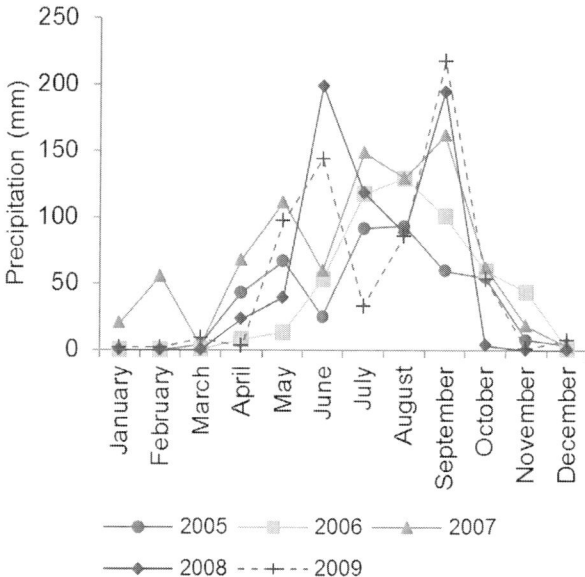

Figure 12. Precipitation records for the Tepeaca district for 2005-2009

FIGURE 13. EROSION PROBLEMS IN THE PIEDMONT ZONE NEAR THE TOWN OF ACATLAN IN THE TEPEACA VALLEY. PHOTO TAKEN FROM SOUTH TO NORTH.

Pine-oak forest: 2500-3000m elevation	
Abies religiosa (*oyamel*)	*Litsea* sp.
Adiantum capillus-veneris	*Pinus ayacahuite* (*ayacahuite*)
Alnus jorullensis (*aile*)	*Pinus michoacana* (*ocote escobilla*)
Arbutus glandulosa (*madroño*)	*Pinus patula* (*ocote*)
Arbutus glandulosa (*madroño*)	*Prunus capuli*
Baccharis conferta	*Quercus crassifolia* (*algodoncillo*)
Bromus sp. (*zacate*)	*Quercus laurina* (*laurelillo*)
Buddleia cordata	*Salix paradoxa*
Eupatorium glabratum (*escobilla*)	*Salvia* sp.

FIGURE 15. PLANT SPECIES PRESENT IN PINE-OAK FORESTS COMMUNITIES OF THE LA MALINCHE VOLCANO.

FIGURE 14. PANORAMIC VIEW OF THE TECAMACHALCO MOUNTAIN RANGE.

Oak forest: 2200-2500m elevation	
Arbutus xalapensis (*madroño*)	*Quercus crassifolia*
Baccharis conferta (*escobilla*)	*Quercus crassipes*
Eupatorium sp.	*Quercus glabrescens* (*encino*)
Juniperus deppeana (*sabino*)	*Quercus obtusata*
Muhlenbergia macroura	*Quercus rugosa* (*encino*)
Pinus leiophylla (*ocote chino*)	*Salvia elegans* (*mirto*)

FIGURE 16. PLANT SPECIES PRESENT IN OAK FORESTS COMMUNITIES OF THE ACAJETE RANGE.

average between 16-18°C with low precipitation and high evaporation rates (Figure 17). Such conditions force farmers to rely on more precocious plant varieties in order to take advantage of the short time span in which the rains arrive. The eastern portion, starting at Quecholac and Tenango, is a transition to the drier and cooler areas of the Palmar de Bravo Valley. Heading north and west, altitude is 100 to 200m higher and temperature gradually becomes cooler with mean temperatures fluctuating around 15-16°C in the Tepeaca and Acatlán area. Evaporation-transpiration rates are also higher in the southern valley than in the north (Figure 18). Therefore, although precipitation might be adequate during any given year, evaporation generally tends to surpass precipitation, which in areas with shallow or sandy soils with poor moisture retention decreases water availability for plants.

Tectonics, local geology, water erosion, and in some cases deforestation, have produced a landscape were intermittent water streams appear in crags and gorges during the rainy season. Several ephemeral streams originate along the flanks of the La Malinche volcano and the Tezontepec and Cuesta Blanca mountain ranges. These descend southwards to the Tepeaca alluvial plain. Here, the *Barranca del Aguila* or eagle's gorge is a major fault that acts as a large drainage system with a south-southeastern orientation (Medina 2001). It originates on La Malinche and runs south until it reaches the north flank of the Amozoc range. From there, it heads east bordering the range and passes just north of Acatlán and Tepeaca eventually turning south passing Xochiltenango and continuing as a tributary of the Atoyac River. During the rainy season, the flow of water can be considerable and in the past it was used to irrigate

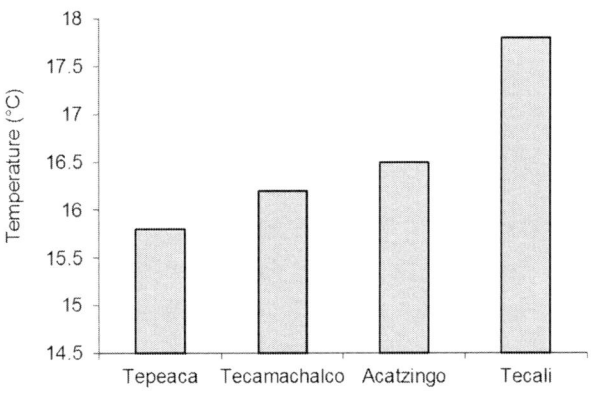

FIGURE 17. MEAN TEMPERATURES FOR FOUR STATIONS IN THE TEPEACA VALLEY

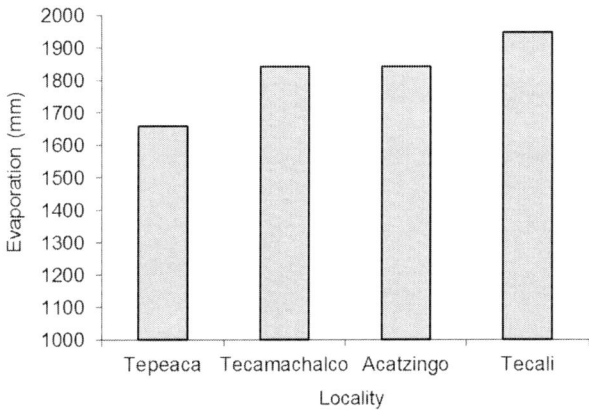

FIGURE 18. MEAN EVAPORATION (1971-2000) RATES IN FOUR LOCALITIES IN THE TEPEACA VALLEY

Massive clayish Vertisols are found near Cuauhtinchan and in mountain ranges east of the Acajete plateau. Cambisols comprise most of the north-central portion of the valley around the towns of Obregón, Tenextepetl, Macuilá, Villanueva, Santa Ursula and Tenango. Some rocky Litosols and Rendzina soils rich in organic material but prone to erosion, are found in the west around Acatlán and the Amozoc range.

Vegetation contrasts are noted as one travels from north to south. Xerophilous bush and scrublands and pasture land vegetation types such as mesquite (*Prosopis laevigata*), izote (*Yucca periculosa*), acacias (*Acacia bilimekii*), maguey (*Agave* spp.), Beucarnea gracilis, Hechtia roseana, and the induced southamerican *pirul* tree (*Schinus molle*) are found in the central, east and southern portion of the valley; this is a transition zone to even more arid areas of southern Puebla (Rzedowski 2006: 116-117,224,259, 265-266). As mentioned before, the north Acajete plateau has oak forest communities. In addition, the lower piedmont areas of the Villanueva and Tenango vicinity present large stands of nopal and *tuna* (prickly pear, *Opuntia* spp.) (Figure 19).

The Llanos de San Juan

The Llanos de San Juan region is separated from the Tepeaca valley by the Tezontepec and Cuesta Blanca fields along its lower banks. The areas most influenced by this system are those in the vicinity of Xochiltenango and Tepeaca.

According to Mexico's *Secretaría de Agricultura, Ganadería, Desarrollo Rural, Pesca y Alimentación* (SAGARPA), around 90% of local agriculture is based on rainfed systems due to the lack of surface water and permanent streams. The surface destined for rainfall agriculture varies between years within the central Puebla region. The reader may wish to access www.oeidrus-df.gob.mx/aagricola_dfe/ientidad/index.jsp for detail information for a particular year. There is, however, an important modern canal system that irrigates portions of the southern valley, which is fed by the Miguel Aleman dam, also known as the Valsequillo dam, just south of Puebla City. Soils inside the valley are of several types (see Figure 10). The southeastern corner, which consists largely of poorly developed Xerosols with some rocky calcareous phases. In the southwest, around Tepeaca, Carpinteros and Zahuatlan, there are Phaeozems rich in organic material and extremely good for agriculture. Phaeozem soils, and Cambisols, are also around Tecali.

FIGURE 19. NOPAL PLANTATION NEAR THE TOWN OF VILLANUEVA LOCATED IN THE NORTH PORTION OF THE TEPEACA VALLEY.

FIGURE 20. PANORAMIC VIEW OF THE LLANOS DE SAN JUAN NEAR THE TOWN OF EL SECO.

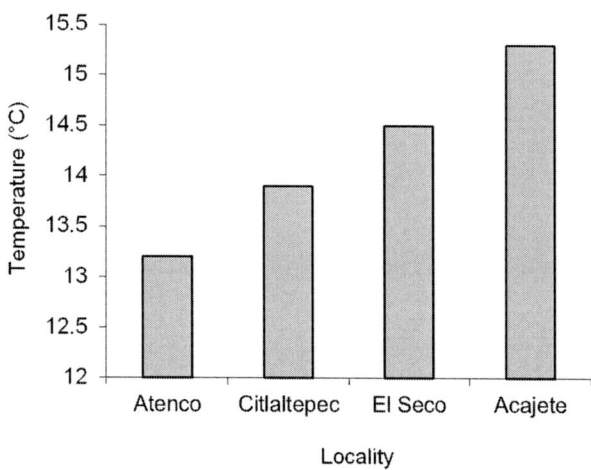

FIGURE 21. MEAN TEMPERATURES FOR FOUR STATIONS IN THE LLANOS DE SAN JUAN REGION

mountain ranges (Figure 20) which range between 2400 and 2700m in elevation (Fernández 1977). Mexico's CNA has registered rain totals of 500-600mm in the Atenco eastern area with a mean annual temperature of 13°C and an elevation of 2450m (Figure 21). The western area, near Zitlaltepec, Tlaxcala and the la Malinche volcano, has elevations that range between 2400 and 2800m with precipitation amounting 800 to 900mm and a mean annual temperature of 13.9°C. In Soltepec, Mazapiltepec, Nopalucan and the El Seco zones, elevations vary between 2400 and 2500m and precipitation reaches 600-700mm with a mean temperature of 14-15°C. Differences in elevations and temperatures allow farmers to plant varieties of maize that ripen in six months in areas where there are deep soils and higher precipitation regimes.

Litosols, Rendzina and Cambisols are found throughout much of the region, especially in the central and western areas (see Figure 10). Fluvisols, Luvisols, Andosols and Regosols are present in the piedmont areas of the La Malinche volcano. The perimeter of the Acajete plateau is composed of Andosols. Phaeozem and Regosols are the main soils in the El Seco vicinity, while the higher elevations around Atenco have Cambisols, Andosols and some Phaeozem soils.

The recurrent wind patterns in central Mexican make the Llanos de San Juan very windy, dry and seriously prone to eolic erosion. The region is well known for its underground water reservoirs fed by glacial melt of the Citlaltepetl volcano glaciers. When economically possible, populations have exploited these underground reservoirs using deep wells extracting the water for agriculture and the production of commercial crops like hybrid maize, alfalfa and potatoes.

Evaporation also plays a crucial role in ground humidity within the study area. According to Mexico's *Comisión Nacional del Agua* (CNA), the eastern portion is considerably more prone to evaporation than the central

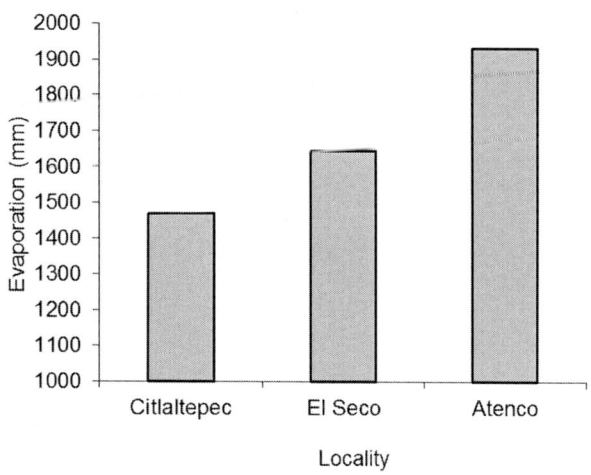

FIGURE 22. MEAN EVAPORATION (1971-2000) IN FOUR LOCALITIES IN THE LLANOS DE SAN JUAN

and western areas (Figure 22). The highest incidence of evaporation is recorded in Atenco and the lowest in the west around the Zitlaltepec area.

The Puebla-Tlaxcala valley

The eastern portion of the Puebla-Tlaxcala valley has a considerably different precipitation regimen than the Tepeaca valley. The *Comisión Nacional del Agua* reports that Amozoc and Xonacatepec receive 800 and 1000mm of rainfall respectively and reflect a climatic regimen influenced by the La Malinche Volcano (Figure 23). Temperature ranges around 15-16°C. Most of the area consists of flat plains or gently sloping piedmont that meets and merges at approximately 2400m. Six-month maize varieties are usually cultivated, except in the higher altitudes above 2700m where longer developing and more frost tolerant varieties are grown (Instituto Nacional de Investigaciones Forestales 1997). High areas between 2300 and 2700m have recurrent frost problems. Fluvisols,

The Natural Setting

FIGURE 23. PANORAMIC VIEW OF THE EASTERN PART OF THE PUEBLA-TLAXCALA VALLEY.

Luvisols, Cambisols, Phaeozem and Andosols are the main soil types and have good agricultural potential (Werner *et al.* 1978). The area around the Amozoc range is covered with oak and pine forests. The valley plains are mostly used for agriculture and contain the modern towns of Nenetzintla, Tlaxco, Tepatlaxco, Acajete and La Joya. As one moves north across the valley towards the volcano, the forests appear again and become denser as altitude rises.

No data on evaporation is available for the towns of Acajete, Xonacatepec or Amozoc. Nonetheless, it is probable that the rates fluctuate between the mean values reported for Zitlaltepec (1400mm) and Puebla (1800mm), that is, of around 1600mm.

This chapter has summarized the natural environmental setting found in the eastern Puebla region. This demonstrates that variation in micro-environmental conditions found here have an important effect on agricultural production of both modern and ancient agricultural populations that resided in the study region.

Chapter 4
Regional History and the Tepeaca *Altepetl*

Tepeaca and the Puebla-Tlaxcala valley during the Postclassic

The Tepeaca *altepetl*, known in prehispanic times as *Tepeyacac Tlayhctic*, was an important state-level polity of the Late Postclassic. The *altepetl* was the main political unit of organization among Late Postclassic societies. It controlled a large territory as well as several populations and ethnic groups. We lack exact information on the size and extent of the Tepeaca *altepetl* boundaries for prehispanic times, but they may have been very similar to those prevalent for the first part of the 16th century AD. The domain probably extended from the towns of Xonacatepec and Amozoc in the western section of the valley to Acultzingo and Napateuctli to the east and southeast areas (Martínez 1984b: 54). The core area of the local lordship covered around 1310km². The limits of the *altepetl* were set to the northwest by the volcano Matlalcueye (today known as La Malinche), and the towns of Nopalucan to the north, Alxoxoca to the northeast, Tenango and Quecholac to the east, Tecamachalco to the southeast, Zacauilotla to the south, Cuauhtinchan and Tecalco (Tecali) to the southwest, and Amozoc and Xonacatepec to the west (see Figure 2).

Many tributary communities lay within the Tepeaca *altepetl*. It is probable that at the onset of the Spanish Conquest, settlements in the region varied both in size and density. Legal documents of the time show that there were many communities in this area. Population estimates are variable and the historical sources are ambiguous. As it is true for other areas of the central Mexican highlands, the number of people living at the time of the conquest may have been high (Sanders 1992). Juan de Torquemada (1969 [1615]) mentions that over 30,000 *vecinos* (neighbors) lived in Tepeaca. As many as 100,000 people may have resided in and around the center. This conclusion is based on the *Relación de Tepeaca* (Molina 1985 [1580]) which explains that more than 60,000 persons died during the epidemic events of the 16th century AD preceding Torquemada's estimates. A much higher figure was proposed by Cook (1996: 113) who calculates 250,000 persons living in the Tepeaca region with a population density of 480 persons/km².

Archaeological data collected by the PAT project, directed by James Sheehy and Kenneth Hirth (Sheehy *et al.* 1995; Sheehy *et al.* 1997), and settlement patterns reconstructions made by Heath Anderson (2009), also support the presence of an important number of settlements for the Late Postclassic. Unfortunately, the bulk of archaeological work has concentrated in the southeastern, eastern and western parts of the state of Puebla (Plunket and Uruñuela 2005) and, aside from the PAT project, little work has been carried out in the central portion which includes the Tepeaca region. Some data have been published (Castanzo and Anderson 2003; Castanzo and Sheehy 2004), but most pertains to the early periods and is condensed into technical reports and theses (Anderson 2009; Castanzo 2002; Maldonado 1997; Medina 2001). Archaeologically, the Postclassic is largely understudied and most of our knowledge on societies of this period comes from historical accounts and legal documents.

We know that larger populations, sometimes referred to as *cabeceras* (head towns) by Spanish chroniclers, were present and recorded in historical accounts. In particular, the towns of Acatzingo and Oztoticpac had a prominent position within the Tepeaca *altepetl* and in some respects acted as political entities with independent management. These towns possessed the privilege of having their own set of rulers, semi-independent decision-making, their own military organizations, and their own tributary lands with dependent labor. Nonetheless, their territories lacked strict boundaries and they comprised a single continuous geo-political unit with loose frontiers and intertwined territories, producing a patchy land tenure arrangement (Martínez 1984b: 27-44, 53-55). The Suma de Visitas (Cook 1996: 112) stipulates that Tepeaca 'Linda con Azumba y con Cachuela (Quecholac) y Tecachalco (Tecamachalco) y con Tecalco (Tecali) y estos pueblos están entremetidos unos con otros y casas con casas' [borders with Azumba and with Quecholac and Tecamachalco and with Tecalco, and these towns are intertwined one with the other and houses with houses] (my translation).

Regional history and the origins of the Tepeaca altepetl

Tepeaca developed during a time characterized by important population migrations, conflict episodes, political instability and the formation of factions and alliances between different regional *altepemeh* (Figure 24). The post-Toltec migrations and Chichimec conquests were central to the political shifts in power between the region's rival *altepemeh* and a strong stimulus for the sharp demographic changes. Between the 12th and 15th centuries numerous wars occurred between the communities of Chalco, Cholula, Cuauhtinchan, Huexotzingo, Oztoticpac, Tecamachalco, Tlaxcala, and Totimehuacan who sought to establish regional hegemony in the Puebla-Tlaxcala valley. The motivations for these conflicts varied, but most were centered on the control of land and labor and major trade routes (Dyckerhoff 1978; Gámez 2003; Martínez 1994b). These dynamic conditions had a profound impact on the socio-political configuration of the Puebla-Tlaxcala valley generating sharp ethnic and political diversification

Date	Event	See
12th century	Toltec-Chichimec vs. Olmeca Xicallanca wars in Cholula	Kirchhoff et al. (1976)
12th century	Chichimec conquests in central Puebla	Kirchhoff et al. (1976)
1178-1182 A.D.	Foundation of Tepeaca and the conquest of the northern territories	Martínez (1984b)
1228 A.D.	Huexotzingo-Tepeticpac "civil wars" of 1228 A.D.	Muñoz Camargo (1988 [1580])
13th century	Cholula-Acolhua wars	Kirchhoff et al. (1976)
1398 A.D.	1398 Conquest of Cuauhtinchan	Reyes (1988a, 1988b)
15th century	Huehuecuauhquechollan vs. Huexotzingo and Calpan wars	Dyckerhoff (1978), Plunket (1990)
1441 A.D.	Conquest of the Mixteca-Popoloca town of Oztoticpac	Kirchhoff et al. (1976), Solís (1992), Velázquez (1945)
1458-1466 A.D.	Land disputes between Cuauhtinchan and Tepeaca	Reyes (1988a, 1988b)
1466 A.D.	Conquest of Tepeyacac	Martínez (1994b)

FIGURE 24. MAJOR CONFLICT EVENTS IN CENTRAL PUEBLA BETWEEN THE 12TH AND 15TH CENTURIES.

and an unstable and an environment of mistrust between neighboring communities.

According to the *Historia Tolteca Chichimeca* (Kirchhoff et al. 1976: [319, 320]), the area where Tepeaca was founded had previously been occupied by a variety of ethnic groups including the cozoteca or citeca-cozoteca, olmeca-xicallanca, poctecatl, tlacuiloteca, tohtepehua and ayauhteca. In AD 1178 Chichimec migrants known as the colhuaque-tlamayoca, which probably came from Huexotzingo, were persuaded by Cuauhtinchan to colonize the mountain chain known as the *Sierra de Tepeaca* and other nearby areas. In AD 1182, a second wave of migrants known as the tepeyacatlaca entered and was established in the same area.

Cuauhtinchan probably encouraged the colonization of Tepeaca to attain its expansionist interests and control the eastern and northern sections of the valley. The establishment of groups of allies served as a method for subduing and diminishing the presence of already established native populations and as a bastion for further conquests. Thus for most of its early history Tepeaca was a tributary of the Cuauhtinchan *altepetl* and under its political control.

Over the course of the Late Postclassic Tepeaca became a prominent and powerful center. In part, it gained prominence owing to the several conquest expeditions it undertook resulting in the subjugation of the northern territories. By doing so, it extended its domain from Acatzingo to Nopalucan and into the area surrounding the Matlalcueyetl volcano. These local conquests reflect the regional growth in power.

Late in the 14th century AD, a series of internal conflicts occurred in Cuauhtinchan. This was the result of a division between nobles from the nearby town of Tecali and a series of conflicts between them. In order to end the disputes Tecali asked for the intervention of Mexico Tlatelolco, the twin sister of Mexico-Tenochtitlan, who already had an important presence in southern Puebla (Plunket 1990). In AD 1398 Cuauhtinchan was conquered and the Chichimec linage of Teuhctlecozauhqui was finally overthrown after ruling for more than 200 years. From this moment, the control of the *altepetl* was transferred to the Pinome group of mixtec-popoloca origin.

After the conquest of Cuauhtinchan, Tepeaca seems to have initiated a series of actions aimed at obtaining independence. In the 15th century AD, the nobility of Tepeaca and Cuauhtinchan engaged in a series of conflicts trying to attain control over the land and human resources of the region. Around 1457, the land disputes intensified ending in war which resulted in the military defeat of the Cuauhtinchan and Tecalco rulers in 1457-58. The Pinome were thrown out of the region and had to seek asylum in the southern *altepetl* of Matlactzinco. For the next two years neither community had complete control over the region's commoner labor pool and land resources. However, between 1459-60 to 1466, Tepeaca made use of the provincial Cuauhtinchan lands and collected tribute from the local populations. It is at this time when Acatzingo became politically dependent and a tributary of Tepeaca (Martínez 1984b: 46).

From exile, the Pinome nobility solicited military and diplomatic aid from the Mexica Empire. For the former, it was a means to regain control of their lost possessions.

For the latter, this became a good opportunity and excuse to take over an important commercial route between the Basin of México and the eastern and southeastern lowlands of the Gulf coast and the Pacific Ocean. Historical records indicate that the Mexica intervention occurred in AD 1466. What the texts do not clarify is whether the Mexica ruler in charge at this time was Moteuhcuzoma Ilhuicamina, as Diego Duran states it, or Axayacatzin as it is said in a 1546-1547 historic manuscript from Tepeaca (Martínez 1984b: 48). In all probability it was Moteuhcuzoma, who according to Alva Ixtlilxochitl (Alva Ixtlilxochitl 1997: Ch. XLVI, p.141) died in 1469 and may have been very old and near his final years, but was still the ruler of the Mexica. We know that Axayacatzin became the next successor, and it is most likely that by the time of the Tepeaca incursion he was the general in charge of the operation.

Historical records explain the reasons behind the conquest of Tepeaca, although they appear somewhat confusing and contradictory. For one, the Chronicler Fray Diego Durán writes that the Mexica justified the invasion because of the assassination of merchants while passing through Tepeaca. The Mexica considered this as an open defiance on part of the Tepeaca leaders to the Aztec's imperialistic goals. For them, it represented a direct summoning to war (Alvarado Tezozomoc 1944). The second version is related to the previously mentioned usurpation of the Cuauhtinchan lands by Tepeaca and the following conflict between both *altepemeh*, which ended in the eviction of the Cuauhtinchan nobility.

In any case, the conquest of Tepeaca seems to have been quick and easy with virtually no military response. Apparently, Tepeaca was not aware of any Mexica incursion or expecting their attack. Upon arrival, the Mexica sent spies to evaluate the enemy's forces. They were surprised —if indeed this is true— to find no evidence for the defense of the city and no indication that the Tepeaca people were preparing themselves for war. Whatever the case may be, the head town of Tepeaca and the neighboring center of Acatzingo were overpowered in a day or two and became yet another province forced to pay tribute to the Mexica in goods and service (mainly military), where Tepeaca was the only tributary province requested to provide war captives as tribute payment.

At the same time, Moteuhcuzoma ordered the creation of a local marketplace in which merchants could come and sell their products:

> Your lord the king also orders that all those outsiders who wish to may go live in your province. They must be given land where they can dwell. Thus, your city of Tepeaca will be made greater with these people from other areas. The king also wishes that a great marketplace be built in Tepeaca so that all the merchants in the land may trade there on an appointed day. In this market there will be sold rich cloth of all kinds, precious stones and jewels, feather work of different colors, gold, silver, and other metals, the skins of animals such as jaguars, ocelots, and pumas, cacao, fine breechcloth, and sandals. All this out lord King Motecuhzoma orders you to carry out 159 (Duran 2006 [1579]: Ch. XVIII) (my translation).

As odd as it may seem, the conquest of Tepeaca probably brought about more advantages than drawbacks to its rulers. Most of the power and government over the region, which was formerly under Cuauhtinchan's domain, was transferred to Tepeaca. It also stabilized a politically and military volatile area and gave Tepeaca economic prominence by becoming the Mexica's regional calpixcazgo (tax collector center), home to large numbers of permanent and passing Mexica settlers, and a strong trade and economic center.

One of the most profound implications of this conquest was the rearrangement of the region's political territories and frontiers. The area was restructured to fit the tribute collecting demands of the Mexica and, omitting the previously recognized boundaries, the land was divided into five major territories between Tepeaca, Tecalco (Tecali), Tecamachalco, Quecholac and Cuauhtinchan. These communities, together with the other major 17 towns in the region formed a major tributary province (Berdan and Anawalt 1992). Nonetheless, even though the Mexica established regional hegemony, territorial disputes between neighboring *altepemeh* continued. In the case of Tepeaca and Cuauhtinchan, these disputes continued well into the Early Colonial period over the control of lands located within the vicinity of Acatzingo (Iglesias 2000). At the regional level, the event had severe repercussions for the Puebla-Tlaxcala Valley. Two major antagonistic zones arouse: one comprised of those populations conquered by the triple alliance and located in the eastern part of the valley; the other consisting of the independent *altepemeh* of Cholula, Totimehuacan, Huejotzingo and Tlaxcala in the western section of the valley (Berdan 1994: 301-302) (see Figure 1).

Tepeaca continued to be a prominent community during the Colonial era. Tepeaca is first mentioned during the events of the Conquest in 1520 when the Spanish army, aided by thousands of indigenous warriors from Tlaxcala and Zempoala, subjugated the town and renamed it as *Segura de la Frontera* (Cortés 1992; see also pictorial scenes of the Conquest in the *Lienzo de Tlaxcala* in Chavero 1979). For a short period, it acted as the main Spanish outpost in the region. By this time, the Tepeaca marketplace was already one of the most important commercial centers in the area. Exchange occurred both at the local level and pan-regional levels and involved a wide diversity of products introduced from several regions including the south and southeast lowlands. The marketplace survived the Colonial period and continued to thrive up to modern times (Seele *et al.* 1983). The reason may be due to both the tenacity of the local populations to retain their indigenous traditions and the value and importance of markets for the support of Spaniards and the general economy of New Spain.

Social structure in Tepeaca and the Puebla-Tlaxcala valley in the 16th century AD

In many areas of central highland Mesoamerica, the organizational backbone of Late Postclassic highland communities was the *altepetl* (pl. *altepemeh*), also called the señorío by the Spanish. The *altepetl* has been the subject of various interpretations, most of them related in a metaphorical and figurative sense. In the 16th century AD the *altepetl* was an indigenous concept widely used in the historical and legal documents written in Nahuatl. It was James Lockhart (1992) who proposed to use the term as an analytical category for it refers both to a territory and the organization of individuals within its domains.

The *altepetl* was used synonymously to talk about a territorial state and the people settled in its patrimonial territory. It has also been described simply as a water-filled mountain because it recurrently was represented as an image composed by such two elements (Florescano 2006). Smith and Hodge (1994) think that it alludes to the juridical claims of land and water, the basic necessities for human sustenance. For Bernardo García (1998: 62-63), the word literally means 'water-hill' which would be a reflection of the surrounding physical environment, pointing out its territorial expression. At a higher level, the *altepetl* would have had, by definition, a central ruler or tlahtoani, a central market place, a temple dedicated to the main deity, and other civic-ceremonial entities.

Yet, virtually everyone recognizes that the term is composed of two words which draw attention to two natural elements: *yn atl yn tepetl*, literally 'the water' and 'the hill'. This is what Mexican scholars have called a '*disfrasismo*' in which two words are united to produce a third one. Literally it may be translated as 'water filled mountain', but in the metaphorical sense it refers to a town or a population because it points out the two main necessities for human sustenance: land and water.

Postclassic *altepemeh* were small political entities structured like archaic states, although with considerable differences. The *altepetl* reflects the political and communal arrangements of Postclassic indigenous populations and is useful for analyzing their social composition. There were no particular territorial limits for a number of subject towns and people needed to define an *altepetl*. On the contrary, each one had different, sometimes contrasting, territorial extensions; these variations applied also to their internal structure and rank order.

Lockhart (1992) argues that there was a more complex social-political organization than the simple *altepetl*; he called this the 'complex *altepetl*'. These larger units were in effect, expanded *altepetl* formed through two different processes: 1) the alliance of various *altepemeh* that seemed to have grown through the progressive subdivision and separation of communities from their previous group (e.g., Tlaxcala and Tenochtitlan) or 2) by the fusion of some of them into a single polity due to their common historical experience that generated a sense of common ethnic origin and affiliation (e.g., Chalco and Texcoco). In these complex states, each *altepetl* performed equal roles and had the same prominence, yet they were clearly separate entities linked to one another using principles of hierarchical organization and rotating leadership responsibilities within the *altepetl*.

Several other social subdivisions may have existed in the *altepetl* that have not yet been documented by researchers. For example, Schroeder (1991) has detected in the writings of Chimalpahin the use of the term tlayácatl for minor sub-divisions or ramifications of the *altepetl*; the Spaniards called these divisions the parcialidad or parte. Also, Lockhart (1992) mentions that in a mid-16th century AD Nahuatl census from Tlaxcallan section divisions of the *altepetl* are termed tequitl, which although there is no clear criteria for the use of such divisions, may be that they are simply groupings created for purpose of the census.

For each *altepetl* there was one supreme ruler controlling the populations and lands embedded into its political realm. Occasionally, other conquered or otherwise dominated *altepemeh* might form part of it (Hirth 2003). Because the *altepetl* was an intertwined set of land tenure arrangements and political subdivisions, there was no clear distinction between the core of the settlement and its periphery. Instead, the landscape was dominated by series of towns interspersed with their agricultural fields, giving the appearance of a disperse settlement pattern rather than a nucleated one. Therefore, although *altepetl* might share similarities in terms of social organization with concepts such as 'city-state', this term is not useful for describing or modeling the details of the Postclassic settlement patterns or socio-economic arrangements. In the same way, European conceptions such as the 'city' are not apt for describing and analyzing the prehispanic urban-rural settlements (Gutierrez 2003; Marcus and Feinman 1998).

Several basic structural and hierarchical subdivisions were present in most *altepemeh*. Yet, as a cautionary note, it must be said that the social organization described for those from central and western Puebla, as well as in some parts of the basin of Mexico, might be referring to the specific Chichimec view of the *altepetl*. The categories may not necessarily apply to all Chichimec social and ethnic groups living in the area such as the Tecamachalco's Mixtec-Popoloca (Gámez 2003). In addition, the categories, as well as internal political structures between different nahua-chichimec groups, might also be different between different *altepemeh* within the same macro-region. Such differences may account for contrasting forms of political organization including council collective and cooperative governments in the eastern nahua area as opposed to the development of imperial-scale political apparatus in the western portion (e.g., Carballo *et al.* 2012; Fargher *et al.* 2010a; Fargher *et al.* 2011).

The *teccalli* or noble house was the core political institution of the *altepetl* in many areas within the Puebla-Tlaxcala region (Anguiano and Chapa 1982; Carrasco and Broda 1982; Chance 2000; Hicks 2009). In Tepeaca, the *teccalli*, according to Martinez (1984b), was known locally as the *tlahtocayo*. I consider that the term *teccalli* describes in better detail the economic relationship and internal organization of the Tepeaca region populations. Therefore, in order to concord with recent research, I will employ this term instead of that of the *tlahtocayo* which refers basically to the political government or reign of a noble elite sector. The *teccalli* was headed by a high ranking noble (*teuhctli*) and formed a distinctive political entity and the fundamental subunit of the *altepetl*. Each *teccalli* had a crucial role in the economy for it possessed patrimonial lands and communal tributary populations that provided labor, service and goods. For nobles, it became the principal means to access wealth and staple resources. Several *teccalleque* could be present within one *altepetl*, all functioning independently of one another with its own set of authority relations and reciprocity between the highest rank noble and its lesser dependent nobles (*pilli*). Yet, together, they shared the same sense of affiliation to the *altepetl* and were geared ideologically to achieving the same political interests. In Tepeaca, we know of at least four *teccalleque* present at the time of the Spanish Conquest, of which the *teuhctli* of the largest and most powerful one acted as the altepetl's almighty ruler and representative.

This socio-political arrangement generated a highly polarized system of land tenure in which nobles controlled vast extensions of patrimonial lands and a large number of *macehualli* tributary commoners within its domains. The commoner sector was responsible for generating the necessary agricultural food and sustenance base of the whole community. It was composed of the common people, that is, those that were not of noble origin. For the Spaniards they were basically vassals, that is, people subject to payment in return for the use of the land which the lord provided in exchange for their loyalty and military service.

Macehualli were unrelated genealogically to the noble sector. Historical accounts show that war and conquest events were intended to take political control of the region (López 2012) (See Chapter 7). The victors generally retained most of the native populations, usually composed of several ethnic groups, and incorporated them into the economic system as tributaries. As a result, *macehualli* populations generally did not share an ethnic or social affinity with the ruling apparatus. This view was so embedded into indigenous perceptions, that when the Spanish defeated the Tepeaca nobles and took political control over the populations, the *macehualli* sector considered that the payment of tribute should only to be given to the Spanish Crown; they explicitly refused to continue as tributaries of their indigenous nobles (Martínez 1984a: 466-467).

Legal colonial documents from Tepeaca that discuss property trespassing, inheritance and the distribution of working lands, constantly speak about a sector of the population known as *terrazgueros* or *renteros*. They were termed *terrazgueros* because they were obliged to pay a rent in order to have access to the lands of a noble. One document indicates that '…hago presentación de esta memoria y pintura por la qual consta de todas las tierras que el dicho mi menor [Esteban de Mendoza] tiene y posee los maceguales renteros…' (Archivo General de la Nación Tierras, Vol. 2782, exp. 40, f.11r). A second one mentions that '…[Dionisio de Mendoza] tiene sus sementeras que le siembran sus terrazgueros; las quales siembran los principales sus parientes descendientes de su casa y señorío, las quales siembran sus maceguales…' (Archivo General de la Nación Tierras, vol. 2782, exp. 40, ff. 7v-8r). These individuals have often been associated with the landless peasant segment, also referred in Nahuatl as *mayeque* (Carrasco 1989; Zorita 1942). In Tepeaca, all we can state is that *terrazgueros* refer to those people that solicited lands from the *teccalli* as renters through contractual relationships with their noble owner. Commonly, petitions for land were made mainly among the commoner populations. However, other sectors would also solicit land, such as merchants and bricklayers (Martínez 1984a: 447-448, 454-455). It's not clear why some individuals needed additional lands. Population pressure or scarcity of good lands may have been factors that account for this problem. This is especially relevant if we consider the high number of inhabitants in the area prior to the conquest and the strong political conflicts between rival *altepemeh* over large uninhabited areas in the boundary zones destined for war purposes (e.g., Plunket and Uruñuela 1994). However, the excessive control of the local nobles over the *altepetl* territories has been stated as the primary cause for land scarcity among the commoner sector. It has been speculated (Reyes 1988a: 114) that the appropriation of the *macehualli* lands often through violent means by the nobles, resulted in the proliferation of individuals that lack lands. Consequently, many of these individuals were converted into simple *terrazgueros*. However, even though the legal documents from Tepeaca indicate that nobles possessed a large quantity of lands, it is clear that some sectors of the *macehualli* population had access to patrimonial plots (Carrasco 1973: 32).

A final category was the *calpulli*. This was a distinctive corporative institution that is mentioned in the Tepeaca historical documents and recurrently in the *Historia Tolteca Chichimeca*, the most important historical document of Late Postclassic Chichimec history. Most of our knowledge about *calpulli* composition has been extracted from the writings of the 16th century AD oidor Alonso de Zorita (1942) who details its social and economic organizational structure. He reports that this organization was present in both the Puebla-Tlaxcala area and in the Basin of Mexico. Nevertheless, some scholars (Carrasco 1997; Hicks 1982: 224) think that it might actually be a Puebla-Tlaxcala phenomenon because Zorita's descriptions might have been based on observations by Francisco de las Navas, a cleric that lived and took care of populations at Tecamachalco, Tepeaca and Cuauhtinchan. Martínez (2001) has even proposed that the *calpulli* was just another regional name for the *teccalli* and that both terms were applied to the indigenous noble house institution.

Calpulli literally means 'big house' or 'large house', probably referring to the kinship affiliation to a particular linage or group. The *calpulli* has been described as being composed of several households with affiliation made through kinship ties, sometimes friendship. Its members were mainly endogamous, had the same patron god, recognized the same line of mythical descent, and had their own telpochcalli (boy's school). They were considered both a military and a |political unit. Each *calpulli* possessed its own set of territorial boundaries and maintained control over internal issues regarding land distribution, temple administration, and vigilance (Aguilar 2006; López Austin 1984: 92-93).

In this chapter, I have described the social structure of the Tepeaca region. The main form of social organization was the *teccalli* system also known as the *tlahtocayo*. The *teccalli* promoted a sharp polarization of land tenure in which elite controlled vast tracts of land while commoners were mainly landless peasants that lived in a noble's estate and paid him tribute in agricultural work, craft production and service.

Chapter 5
Traditional Agriculture in the Study Region

The objective of this chapter is to introduce the reader to the annual agricultural activities performed by farmers within the study region. The focus is on maize, the main staple cultivated under rainfed conditions, and to a lesser degree on other secondary crops. The various tasks involved in the agricultural cycle are described below and are summarized in Figure 25. Making an efficient use of the available technology and the cultivated plants depends largely on the farmer's knowledge of his natural environment and the scheduling and timing of the various stages of cultivation. It is the conjunction of the techniques of cultivation, the labor investment, and the environment variables that make the agricultural cycle tremendously complex.

Stages and work in the agricultural cycle

The rainy summer season is a good time to grow staple crops. Maize is the most important crop for household subsistence. Landraces most frequently cultivated within the study region are cónico, arrocillo, and cacahuacintle, while in the southern portion of the Tepeaca valley cónico, cacahuacintle, pepitilla and cuapeño are the most used (for landrace descriptions see Benz 1988; Wellhausen *et al.* 1952). These are the major racial divisions, but there are more localized varieties that differ from one another in color and overall morphology. They are the result of human manipulation and adaptation to diverse local environments. According to Mexico's SAGARPA, there may be more than 300 varieties of maize just within the State of Puebla.

Secondary plants to maize are grown in both the main fields and in houselots. These plants include climbing beans (*Phaseolus vulgaris*), *ayocote* beans (*Phaseolus coccineus*), green tomatoes (*Physalis philadelphica*), red tomatoes (*Solanum lycopersicum*), squashes (*Cucurbita pepo* and *C. radicans*), and chili peppers (*Capsicum annuum*). In areas where winter cultivation is possible — mostly by irrigation from deep water well— smallholders grow cash crops like peppers, lettuce, onions, parsley, and many types of flowers that represent an important income in the fall season when sold during festivities such as Independence day (*Día de la Independencia*) and the Day of the Dead (*Día de Muertos*). Plants that are grown near agricultural fields include *nopal* (cactus and prickly pear —*Opuntia* spp.), *pitaya* and *pitahaya* (dragon fruit cactus —*Hylocereus* spp.), *maguey* (*Agave americana*) and fruit trees like *capulín* (*Prunus capulli*), tejocote (Mexican hawthorn —*Crataegus mexicana*) and the European introduced peach (*Prunus persica*).

To a person unacquainted with Mexican traditional agriculture a visit to the countryside can be confusing. Only a few farmers can be seen performing their quotidian chores. The reason is that in traditional Mexican rainfed systems, agricultural labor is often quite sporadic. Cultivation stages are spaced between one another by weeks and it generally only takes about three days to accomplish them. Therefore, the farmer does not need to be in his field on a day-by-day basis. In addition, farmers time their tasks according to personal observations on climatic phenomena, and in relation to the economic activities in which they are involved. Except for jobs that require workgroups, everyone works on their own according to personal time managing and convenience. Therefore, throughout most of the year large numbers of workers are absent in the fields.

Field preparation

Field preparations usually start between January and February, but can begin as early as November-December. The first labor is the barbecho, which consists of turning and breaking up the soil with deep plowing and to leave the field fallow until initial sowing. Fields are given several passes using a two-sided plow pulled by two animals (horses or mules). The soil is loosened and softened to promote optimal seed germination and early plant development. Plant residues from the past harvest are incorporated into the soil and function as fertilizer. Deep plowing also provides sufficient ventilation to the upper portion of the soil. Water from the sporadic winter rains can be retained in the sub-soil providing moisture for future plantations. In addition, pests (Figure 26) are adequately controlled because larvae and eggs are killed by exposure to cold temperatures or consumed by natural predators like birds.

On average, it takes around 10 hours to finish the barbecho for one hectare. Farmers will try to do the job in one day when working in distant fields, but they will usually do it in two days, preferably during the early morning hours. Weeding (*deshierbe*) takes place before, or at the time of the barbecho. Weeds are removed and burned, and the ashes incorporated to the field as fertilizer. Weeding takes approximately one to two human working days of eight hours/day for one hectare of land. Next, the field is leveled using a mold board (*rastra* or *rastrillo*) to provide a soft bed for plants. In its simplest form, it is an implement made out of tree branches tied to the sides of an animal. Two passes are enough to level the field and it takes about 4 hours to finish one hectare. When using a tractor, a timber is tied to the back and passed around the field taking only a couple of hours to finish the job. Once the field is leveled, the land is left to fallow for three to four months until the start of the rainy season.

Activity	Description	Time
Primer deshierbe	First weeding: dismounting the field and burning up weeds	Between December and February. Before the *barbecho*
Barbecho	Deep plowing of the field to loosen soil, ventilate the field, promote humidity retention and control insect pests	A few days after initial weeding
Abonar	Manure fertilization: application of animal manure to fields	Usually in the winter season and simultaneously with the *barbecho*
Rastreo	Leveling the field with a *rastrillo* or mold board	Immidiately after barbecho
Selección de semillas	Selecting seeds from personal storage or acquiring it from a third party	Between April and June
Surcada	Plowing the field to make linear furrows for sowing	Between April and June
Siembra	Initial sowing	Right after the *surcada*
Resiembra	Re-sowing spaces with poor plant appearance	15-20 days after sowing
Primera labor	Weeding and loosening the soil with the plow. One pass is made.	Around 20 days after sowing. Plants are 15-20 cm in height
Segunda labor	Weeding and loosening the soil with the plow. Two passes are made.	When plants are around 40-50 cm in height
Tercera labor	Hilling plants with the plow	When plants are around 100 cm in height
Fertilización	Chemical fertilization with nitrogen, phosphorus, potassium	Concomitant or slightly after the *tercera labor*
Segundo deshierbe	Second weeding to avoid plant competition	Four months after initial sowing
Recorte de espiga	Removal of top portion of maize stalks used as animal fodder	Around five months after initial sowing
Cosecha o pixca	Harvest	When cobs are dried up around 14% to 18% humidity content (after 180 days for six month varieties and 130 days for four month varieties)
Secado y almacenamiento	Sun drying of maize ears at the household compound and its storage	After the harvest

FIGURE 25. ANNUAL AGRICULTURAL ACTIVITIES FOR MAIZE CULTIVATION WITHIN THE STUDY REGION.

Furrowing and sowing

Furrowing and sowing start right after the first summer rains. Fields are plowed again to form furrows oriented in reference to the longest portion of the parcel. In hilly areas, farmers place furrows perpendicular to the slope and following the natural contour of the terrain. Trees and maguey rows are sometimes set along field borders to reduce soil erosion. The furrow beds are made with a one-sided plow pulled by two traction animals. The work requires around 5 to 7 hours for a hectare, depending on the farmer's experience, the strength of the animals, and the texture of the soil. Soil texture is a particularly important energetic and time management variable. Clayish soils become extremely hard to work when they are either dry or sticky when saturated with water. They have to be worked at the right intermediate point. Sandy soils are much easier to plow and can be worked on a less restricted time frame. However, they drain water rapidly and need to be sown as soon as the rains begin. Alluvial soils are much better in terms of workability and humidity retention. They give farmers a longer period to work them and require less energy consumption. Xerosols, or arid soils, are poor in water retention and have to be worked immediately after receiving precipitation. Such variability in terms of soil quality force farmers to be aware of seasonal climatic conditions and the particular behavior of their fields. By doing so, they are in a better position to take full advantage of the available resources.

Sometimes time management is a problem for farmers that have various fields dispersed across the landscape. In some of them, it may be necessary to sow *en seco* (on dry land). The idea is to anticipate the coming of rains within a 10-15 day span. This is risky strategy because if rains do not arrive at the right time the grain will rot or be eaten by birds and rodents. These factors will affect total plant density and yields. In some instances, farmers will sow even though

Specie	Common name	English name	Attacks
Oligonychus mexicanus	Araña roja	Spider mites	Sap and foliage
Brachystola spp.	Chapulín	Grasshopper	Foliage
Melanoplus spp.	Chapulín	Grasshopper	Foliage
Sphenarium spp.	Chapulín	Grasshopper	Foliage
Dalbulus maidis	Chicharritas	Corn leafhopper	Sap and leafs, produces undergrowth
Empoasca kraemeri	Chicharritas	Leafhopper	Sap and leafs, produces undergrowth
Chelinidea tabulata	Chinche gris del nopal	Cactus/prickly pear bug	Nopal sap
Dactylopius indicus	Cochinilla or grana	Cochineal	Nopal leafs and prickly pear
Epilachna varivestis	Conchuela	Mexican bean beetle	Bean foliage
Diabrotica spp.	Doradilla	Corn Rootworm	Roots and stem
Macrodactylus spp.	Frailecillo	Bettles	Roots
Phyllophaga spp.	Gallina ciega	White grubs	Roots
Keiferia lycopersicella	Gusano alfiler	Tomato Pinworm	Tomato foliage and fruits
Laniifera cyclades	Gusano blanco de nopal	White grub of nopal	Nopal leafs
Spodoptera frugiperda	Gusano cogollero	Moth, Armyworm	Maize stem, ears and kernels
Agriotes sp.	Gusano de alambre	Wireworm	Stem, seeds and shoots
Melanotus sp.	Gusano de alambre	Wireworm	Stem, seeds and shoots
Heliothis spp.	Gusano del fruto	Budworm	Tomato fruits
Pseudaletia unipuncta	Gusano soldado	Moth, Armyworm	Foliage
Liriomyza spp.	Minador de la hoja	Leafminers	Foliage
Xenochalepus signaticollis	Minador de la hoja de frijol	Leafminers	Foliage
Trialeurodes sp.	Mosquita blanca	Whitefly	Fruits and foliage
Cactophagus spinolae	Picudo barrenador del nopal	Cactus weevil	Nopal leafs
Apion godmani	Picudo del ejote	Pod weevil	Bean foliage, flowers and developing seeds
Geraeus senilis	Picudos del maíz	Maize billbug	Tender maize leafs, stem, ears and spikelets
Nicentrites testaceipes	Picudos del maíz	Maize billbug	Tender maize leafs, stem, ears and spikelets
Epitrix cucumeris	Pulga saltona	Flea beetles	Foliage
Myzus persicae	Pulgón del chile	Green peach aphid	Chili, pepper fruit
Rhopalosiphum maidis	Pulgón del cogollo	Corn Leaf Aphid	foliage, produces undergrowth
Toxoptera aurantii	Pulgones	Black Citrus Aphid	Sap and developing foliage

FIGURE 26. COMMON INSECT PEST THAT AFFECT CULTIGENS.

rains are delayed for a month or more from the expected time. Others sow at a fixed date due to time constraints. It is common for farmers to have alternative jobs or be involved in other economic activities. Consequently, dry sowing is a rare practice.

Furrow width will depend on the type of crop to be sowed. When monocropping techniques are used, a ridge distance of 70cm is considered appropriate for maize. Conversely, 50 to 60cm is used for planting ground beans. If maize is intercropped with climbing beans (*Phaseolus vulgaris*), ayocote beans (*Phaseolus coccineus*), or squash (*Cucurbita pepo* or *C. radicans*), a distance of 70cm is also the best choice. While one person prepares the furrow bed, a second one —frequently a women— walks behind the plow carrying the seeds in one or two small bags and drops three to four kernels at a pace distance (60-70cm). Maize kernels are then covered up with a thin layer of dirt using the feet. Before sowing, decisions are made regarding how planting will be done and which varieties will be used. When intercropping, one or two beans are deposited every two steps apart and two or three squash seeds 10m apart. Most farmers avoid sowing climbing beans and squashes together because they entangle and strangle maize plants, thereby reducing total yields. Maize varieties that are more vigorous (i.e., have more strength to sprout) are used in areas with poor retention of soil humidity because kernels need to be put deeper in the ground (10cm).

Farmers commonly possess some sort of livestock or working animals such as horses, donkeys, cows, sheep and goats. Therefore, they prefer to narrow the furrow width to as little as 50cm in order to produce more plant carcass at harvest for use as animal fodder. Others sow maize densely on the field's perimeter, especially when the plot is located near a road. This acts to protect crops because it provides free fodder to animals and discourages them from going deeper into the field.

Farmers commonly plant several lines of maguey along the field's perimeter. This serves as a boundary delimiter, but also as a contention wall against soil erosion problems in areas with steep slopes. Maguey plantations are highly priced, The maguey sap, called aguamiel (honey water), can be obtained from the heart of the plant and is used or sold to local *tlachiucualeros* (sap collectors) to produce pulque, a milky viscous looking fermented beverage (Parsons and Parsons 1990). Additionally, the maguey larvae called chinicuil and tecolli (*Hypopta agavis*) and the gusano blanco (white worm —*Acentrocneme hesperiaris*) can be collected during the rainy season and are sold as a special food in restaurants representing extra income for households (personal observation).

Sowing is done by two adults. Adolescents and children also take part in these chores as part of their introduction to agriculture. One hectare can be sowed in one to two days (10-16 human hours) depending again on the experience of the laborers and the workability of the soil.

Animal manure fertilization

Animal manure fertilization was not an activity carried out by prehispanic populations because they lacked European traction animals such as cows, horses, sheep and goats. Unfortunately, there is no efficient way in which to factor out this input when estimating ancient agricultural productivity. However, we know that indigenous agriculturalists carried out various strategies to enhance soil quality by inputting products rich in nutrients such as night soil (*tlalauiac*), rotten wood (*tlazotlalli*), forest humus (*quauhtlalli*), and house refuse (Rojas 1988). Modern ethnographic data shows that similar methods for soil improvement are common among modern Mexican agriculturalists (Gliessman 2000). This suggests that prehispanic natural fertilization methods might have been an excellent source of nutrients for plants that probably equated animal manure in terms of quality. Therefore, animal manure application strategies can be used as a proxy for understanding the amount of labor investment and the timing of fertilization activities that would have taken place in ancient times.

Fertilization with animal manure is done during the winter season, a month or two after the harvest concomitant with the barbecho. Animal manure is considered better than chemical fertilization. Its major disadvantage is that its application is time and energy consuming. It can take a week or more to spread the manure in the field. It is also an expensive method considering that its price can range from $1000 to $2000 Mexican pesos a load (around 5 tons) given that transport costs vary proportionally to the distance from the manure provider to the field, and the type of manure employed (cow, pig, chicken). It takes two loads of manure to adequately fertilize one hectare and it is an excellent source for reestablishing lost nutrients. The benefits only last two cropping cycles, after which a new load needs to be applied. Some consider cow manure to be of better quality and less 'hot' than chicken dung. Pig manure is also considered good, but has the problem of being stinky. Cow manure is easier to obtain locally, while chicken and pig manure is mostly used in areas where there are animal industries to produce it in good supplies like in El Seco, Quecholac, Tochtepec and Tecamachalco.

Animal manure can augment production considerably, but the investment costs are frequently higher than the gains of the harvest. Maize productivity under rainfed conditions and good fertilization is moderately good, with mean yields fluctuating around 1500kg/ha in a typical year (see Chapter 6). Production in an atypical year can be considerably lower. In addition, the Mexican government has fixed the price of staples such as maize and beans in order to keep prices low for the benefit of the poor ($3.00 to $4.00 pesos per kilogram and $7.00 to $10.00 pesos for beans). Thus, it is practically unprofitable to cultivate basic crops and those produced are usually destined for household auto-consumption. Consequently, farmers avoid using animal manure in fields with poor quality soils. They will use any available product including house waste, charcoal ashes, and animal manure obtained from their personal stock.

Seed selection

Seed selection is the result of a series of decisions made throughout the year. Before the start of the new agricultural cycle household members will discuss the advantages and drawbacks of the landraces cultivated during the previous year. Observations are made on the basis of how crops behaved in relation to water availability, local soil conditions, climatic fluctuations, fertilization, pests, and disease. The benefits and drawbacks are pointed out. At some point, women will discern which varieties were better for cooking, preserved better, and had superior texture, color or taste. Men will constantly take out stocks of stored grain to ventilate the product and to check its preservation, separating seeds that are more tolerant to pest and fungi. Therefore, by the start of planting farmers have a good idea of which varieties to plant under the particular environmental characteristics of their fields. Even though farmers select varieties that performed well, sometimes they experiment with those that perform poorly. Seeds of low producing plants from one field will be planted in another field with different soil and environmental conditions. This helps them determine if the varieties had adaptation problems to a particular setting or if they must be definitively discarded.

Once it is decided which maize variety to plant, seeds are gathered from the seed stock. Maize only requires 25 to 30kg of seed from the previous harvest (a ratio of 1:50 to 1:75) to sow one hectare of land. This is an extremely low amount in relation to Old World grains such as wheat or barley which require up to 1/8 of the previous year's total production. Such a small amount of maize produces a strong selection bottleneck because only a few ears are chosen. In order to obtain 25kg of maize, the farmer needs to select less than 300 cobs with an average weight of 100 grams per cob. Given that the density of plant population under rainfed conditions is about 30,000 per hectare, the selected cobs represent only 1% of the overall genetic pool. The number of maize cobs on which seed selection occurs is relatively low. Therefore, the farmer has to make an important decision when picking the best available seed.

Most farmers select seeds from the central and bottom portions of the ear. These kernels are considered the best in terms of size, color, shape, robustness, texture and overall appearance. Larger and better shaped kernels are separated from the smaller less attractive ones. However, some less desirable ones are always included in the selection. Farmers prefer to select maize kernels from their personal genetic stocks. Yet, sometimes they trade varieties with other members of the community or acquire them at a local grocery store. High Yield Varieties (HYV) need to be bought from specialized suppliers.

The admixture (*mezcolanza*) of several maize varieties occurs regularly. Maize easily interbreeds, especially when neighboring farmers raise different landraces. The high incidence of cross-pollination complicates the maintenance of pure stands, unless the majority of farmers agree to use the same variety. Most of the time farmers focus on their own crops and avoid wasting time trying to prevent admixture. In fact, farmers promote intentional cross-pollination between plants. Information constantly flows between farmers regarding their cultivated varieties and how did they performed. Many try to improve the performance of a maize stock under a particular land type or environmental condition. Varieties that performed poorly are discarded within a year or so, depending on the farmer's determination to stabilize them to local conditions. When transferring plants from one region to another, maize may develop poorly or not grow at all. At other times, transported varieties grow profusely and yet are barren. Thus, two or more maize varieties are commonly sown in the same field. Such a strategy has two advantages. First, it gradually introduces new varieties without risking the entire crop production, and secondly it promotes cross-pollination that leads to the emergence of better adapted landraces. One farmer told me that he wanted to have blue tortillas and hence he planted the entire field with blue maize. In the past years, he had planted white, blue and red varieties together in the same field because he wished to produce *pinto* or ears with multi colored kernels.

Re-sowing

Re-sowing is performed around 20 days after the initial sowing. Gaps created by climatic and predator affectations are re-planted. It is generally a simple task and will take one day for one person to finish the job.

Primera, segunda and tercera labor (first, second and third labor)

Three more plowing's are needed throughout the agricultural cycle besides the barbecho. The first is made when the plant has around 20cm tall. The idea is to loosen the soil and provide better conditions for the plant to growth. The second is made when the plant reaches 40 to 50cm in height in order to loosen the soil and remove competing weeds. Both the first and second plowings are made with a single sided plow and one traction animal. It takes around 10 hours of work, depending on soil workability and weed infestation.

The third plowing is made when the plant reaches one meter in height. Work is done using a 'Z' or two-sided plow pulled by a horse or a mule. The soil is loosened and a hill is created over the root of the plant to provide more stability and lessen the risk of falling. While the farmer plows the terrain, he constantly straightens maize plants and places dirt at their base. He also untangles climbing beans and/or squash from maize plants and rips off competing weeds. The third labor is more laborious and time demanding. For one person, it may take 10 to 16 working hours to finish one hectare.

Second weeding

A second weeding takes place when the plants are around three months old, depending on weed infestation in the field. The work is done manually and takes one to two days for two persons working 8 hours a day to finish one hectare. Weeding considerably reduces plant competition and supplies the crops with better moisture conditions and better nutrient availability. If the farmer does not weed on time, production can fall as much as 50% from the expected yields.

Weeds in the study region are both local and introduced. Some are extremely hard to eradicate such as *coquillo* (*Cyperus esculentus*) and considered highly detrimental for cultivars. Others are useful and are consumed or used as medicine or for rituals. Wild plants consumed include quelites (*Amaranthus hybridus*), epazote (*Chenopodium ambrosioides* and *C. graveolens*), *tomate verde* (green tomato -*Physalis philadelphica*) and *chilacayote* (squash -*Cucurbita radicans*). *Chilacayote* is especially valued for its edible nutritious seeds. Other plants can be used as fodder for farm animals. *Guajolote* (turkeys -*Meleagris gallopavo*) are fed with *polocote* leaves, a wild sunflower (*Helianthus annuus*) that is abundant in the region (Figure 27).

Chemical fertilization

At some point after the third plowing labor, farmers may apply chemical fertilizer to the plants. Fertilizers, mainly nitrogen, phosphorus and potassium, are applied before the plant starts to develop a *xilote* (baby cob) in order to provide nutrients for the adequate development of the ear; *xilote* sprouting takes place around the 120th day after sprouting in a six month maize variety, and 90 days for the fourth month variety. Chemicals can be detrimental to plant growth when water is deficient. If no water is provided to the plant, severe damage can occur to

Specie	Common name	Use	Origin
Atriplex patula	*Quelite*	Food (tender)	Local
Bidens odorate	*Mozoquelite*	Food (tender), tea	Local
Chenopodium album	*Quelite*	Food (tender)	Eurasia
Chenopodium ambrosioides	*Epazote*	Food (tender)	Local
Chenopodium graveolens	*Epazote de zorrillo*	Medicinal	Local
Chenopodium murale	*Quelite*	Food (tender)	Eurasia
Cucurbita radicans	*Chilacayote*	Seeds for food	Local
Datura stramonium	*Toloache*	Hallucinatory	Local
Helianthus annuus	*Polocote*	Food for *guajolotes* (turkeys)	Local
Marrubium vulgare	*Marrubio*	Medicinal	Local
Phalaris minor	*Alpistillo*	Food for captivity birds	Local
Physalis philadelphica	*Tomate, tomatillo*	Food	Local
Plantago major	*Plantago, llantén*	Medicinal	Local
Sambucus mexicana	*Sauco*	Medicinal	Local
Suaeda torreyana	*Romerito, romerillo*	Food	Local

FIGURE 27. SOME USEFUL WEEDS FOUND ON AREAS MODIFIED BY AGRICULTURE.

the root systems killing the plant. A period of 15 days is considered the critical time in which chemical application should be dissolved in water and incorporated to the soil. Therefore, chemical fertilizers and rainfall agriculture are thus a risky combination. Farmers are reluctant to use chemical fertilizers and constantly complain about it. The other problem with this method is it has negative effects on overall land quality. Farmers assert that, in comparison to more natural fertilization like animal manure or household waste, chemicals produce negative effects on soil structure and organic content. Lands that were previously rich in organic content and easy to work after extensive fertilization are now compacted, rough and light colored. Therefore, petroleum based products have a bad reputation for the long-term productivity of soils and are avoided as much as possible. Two persons can apply chemical fertilizer to one hectare in around 8 to 12 hours.

Maize stalk upper portion removal

When the maize ear has fully matured, farmers will remove the upper portion of the maize stalk and use it as fodder for animals. Cutting is made right above the apical ear including the spikelet. Two persons can do the chore for one hectare in around two to four days, depending on plant densities. It is a profitable strategy considering that personal livestock can be fed or the fodder can be sold at local markets. The disadvantage is that the plant residue will not be put into the soil.

Harvest

Deciding when to harvest is a crucial element in food production. Maize has to be harvested at the right time and cannot be done prior to its optimal drying stage of 14% humidity content. If stored with a higher humidity it will rot or be affected by fungi. It left too much time in the field it will be affected by pests and predators, especially birds and squirrels. Additionally, water from late rains can filtrate into the ears augmenting the humidity content. When this occurs, maize must be left to dry for more days. Therefore, even though the farmer tries to avoid these adversities, there is always some percentage of crop loss prior to harvesting.

Harvest is usually performed in workgroups of 4 to 12 persons depending on the size of the field. Traditionally, work was reciprocal and households relied mostly on kinsmen or friends to help out. As they say, 'hoy por mi, mañana por ti' (today for me, tomorrow for you). But in recent times, this reciprocity has turned into an economic relationship. People do not wish to pay back favors. Instead, households have to employ wage labor involving costs of up to $100 pesos per person. They also provide people with at least two meals, refreshments and alcoholic beverages. This obviously implies a higher cost for the field owners. Hiring is needed because many farmers are engaged in other economic activities. There are cases in which farmers have opted to harvest their fields gradually without relying on additional help. Farmers who adopt this strategy are mostly persons older than 60 years that are not, or cannot be, engaged in other time-consuming activities aside from agriculture.

Storage

Storage plays a fundamental role in traditional agriculture. Keeping production safe from predators will result in more food available for the annual needs of the household. Maize comprises the bulk of production and is stored in large spaces. It has to be dried correctly (at around 14% humidity content) in order to avoid fungi development. Most of the drying process takes place at the field. Sometimes, maize has to be harvested when kernels have high humidity content and is taken back to the domestic compound. When this occurs, the cobs need to be laid out to dry. They are set on the roof tops or in an open ground area for several days. During that time special vigilance is needed to guard against bird predation. Making noise, putting children in charge of guarding the maize, or setting up devices like scarecrows can, in some measure, dissuade birds from steeling grain. Anything that helps to scare away birds is welcomed.

If maize is shelled, it can be put in sacks or in containers such as a *petlacalli*, a cylindrical storage deposit made up of woven palm -the same as *petates* or mats- about one and a half meters in height and a meter in diameter (measures may vary), or in plastic barrels. Shelling saves considerable space within the household and is useful if the farmer lacks large facilities. However, some think that the kernels will rot faster and are more prone to be infected by insect pests like *gorgojo* (maize weevil -*Sitophilus zeamais* Motschulsky-) and palomilla (armyworm -*Spodoptera frugiperda*-) when they are removed from their cob. Beans require less space to store and can be done in jars, pots, baskets or small sacks. Beans have a hard pericardium (outer covering) and are more pest resistant than maize.

Maize can also be stored as ears —a cob with kernels and husks— by hanging them in bunches from roof beams inside the house. This method has the advantage of preserving it better because it is well ventilated and kept far from rodent predators. However, only a fraction can be stored in such a way due to limited space availability. If maize is to be kept as ears, they are put in bins like the *cincalli*, a square shaped granary made of pine beams resembling a small corral and topped with a thatched or metal roof (Seele and Tyrakowski 1985), or in sacks inside a special storage room. Ear storage has the disadvantage of being bulky and thus requires larger spaces and special facilities. Squashes are also kept in a special location where they do not interfere with daily activities. Mature squashes are important for seed production rather than pulp. They can be stored for a couple of months until they are opened; seeds are extracted, dried in the sun, and then roasted on a comal (a large flat rounded pan) with salt. Women sell squash seeds as appetizers at local market places. They are an excellent source of oil and an important monetary income for farmer households.

In order to minimize loss to pests, most people will use a mix of storage strategies including the use of chemical aids to suppress insect infestation. *Cincalli* bins are often made out of ocote trees (*Pinus teocote*) which are rich in resin and good for repeling insects. When available, a small amount of tequesquite (mineral salt) is sometimes poured into sacks and other bins and pots, aiding in the control of weevils and armyworm. Traps are set in the storage facilities to catch rodents. Nonetheless, there is no way of completely avoiding pest problems. There is always some loss during the course of the year, which can be highly variable depending on the degree of infestation and the resistance of the maize and beans varieties to pests.

Other aspects of local agriculture

The use of High Yield Varieties (HYV's)

The use of high yielding varieties (HYV's) of grains has penetrated Puebla-Tlaxcala agriculture, especially in areas where the food industry has a high demand for maize products. Maize is consumed in large quantities by cattle and goat milk production. Also, *tortillas* (flat rounded shape food made out of maize) consumption in urban areas has augmented the demand for large scale production. In rural areas, local smallholders can produce enough maize to satisfy their consumption needs. Yet, given that most farmers own small parcels (from 1/16 to 2ha), overall regional production is not enough to supply demand. In some portions of central Puebla, farmers have been able to acquire larger portions of land (10 to 40ha) and are now taking a market economy orientation. It is in these areas where HYV's, irrigation systems, and heavy machinery including tractors are used for large scale production of staples. Nevertheless, they represent a marked minority and consequently important quantities of maize for industry and city consumption are imported from the states of Guanajuato, Jalisco and Sinaloa where large scale agriculture is more common.

In Puebla, maize production using modern fertilization and irrigation techniques can yield eight metric tons/ha or more. Local landraces also yield better under intensive techniques, up to six tons/ha or more (Aceves *et al.* 2002). This is at least five times more productive than traditional rainfed systems. Nonetheless, modern land intensification techniques and the technology associated with it require an equally high investment. For maize, it costs around $15,000 pesos to annually cultivate one hectare of land using modern intensive techniques. Hence, most of the time machinery is rented and bank or governmental loans are required. Given the current economic situation of the country, loans tend to have high interest rates and become a burden for most entrepreneurs. Therefore, on the landscape, there is a clear boundary between subsistence agriculture fields and commercial ones, the later appearing as enclaves in an immense amount of small fragmentary fields. The only link between the two systems is the inevitable cross-pollination that occurs between native landraces and the HYV's due to their imminent closeness.

The Mexican government has tried to introduce HYV's among subsistence farmers with programs such as 'PROGRESA', or through technical support of institutions such as INIFAP and SAGARPA. In part, the objective has been to insert peasants into the national market economy and provide better nutritional status for the people. The strategy has been less successful than expected. A main constraint is the high cost of the technological package associated with HYV's. Farmers do not have enough resources to acquire machinery needed for intensive cultivation of HYV varieties. At other times, commercial varieties are not apt to be cultivated using the local traditional farming strategies which farmers are reluctant to abandon. Instead, they cultivate HYV's in the same way that they do other local landraces. Another problem is the unstable supply of both the improved seeds and the fertilizers packages that the government agencies supply. It is very common that these inputs are irregularly available (after the needed moment). In addition, government aid is not equally available. Those who have a prominent economic position in the local economy, and that are inserted in larger market economy —such as industrial ranchers or milk producers— receive the bulk of the aid, leaving little resources for smallholders. Therefore, subsistence agriculture has had little room for improvement and the trend for farmers is to avoid as much as possible any reliance on governmental support.

What is happening is that many smallholders tend to develop a personal genetic bank based on HYV's and traditional landraces mixtures. By annually sowing, cultivating, harvesting and re-sowing either native varieties or HYV's they have avoided the need to acquire any newly developed technological material —unless it is given for free in the annual cycle that it can be used. Admixture between native maize and commercial HYV's occurs either intentionally or unintentionally. Intentionally because if by any chance they are able to obtain HYV's seeds, these will be planted alongside their traditional stands, thus promoting cross-pollination. Unintentionally due to the proximity of a farmer's field to an industrial maize stand or to someone else's field with HYV's, which also results in genetic mixing. Therefore it is somewhat hard to speak of pure native landraces in the region.

Final remarks regarding modern agriculture in the study region

Strictly speaking, traditional agricultural systems in the study region are the result of a syncretism between indigenous, colonial, and post-colonial practices (González 2003). There is clearly a marked mixture, what Withmore and Turner II (1992: 420) have termed 'mestizo cultivated landscapes'. It is probable that more and very different agricultural systems and crops were present in the past, but most of them became extinct due to the massive depopulation of the central Mexican Highlands after the 16th century AD disease episodes (McCaa 1995). Recently, modern economic and technological developments have gradually become inserted and blended with subsistence agricultural practices giving rise to new forms of land usage. Governmental aid, petroleum based fertilization products, HYV varieties, and large scale irrigation equipment are but a few of the variables that have —and will continue in future years— to affect the way agriculture is practiced among subsistence agriculture smallholders.

Although there is no clear trend as to how these modern changes will influence traditional agriculture systems, for now we can still distinguish indigenous period technological and technique strategies from the colonial and post-colonial ones. The extensive traditional agricultural methods brought in during the colonial period discouraged the more labor intensive techniques used by Mesoamericans such as the personalized rearing of plants described by Early Colonial chroniclers like Fray Bernardino de Sahagún (1963). The plow and furrow system are interesting adaptations to indigenous practices. Plow technology may have been quickly assimilated by indigenous populations. Not only did it allow the farmer to cultivate larger areas with less effort, but it also produced furrows that largely resembled the indigenous manual-made linear ridge mounds called *cuemitl* among nahua groups and *camellones* by the Spaniards. Furrows and *cuemitl* look very much alike and share most of the same functional benefits for regulating temperature, controlling erosion, and distributing water. However, the human energy invested in their elaboration differs considerably. Furrowing one hectare can be done in a couple of days with the aid of animals and the plow. The use of animals, however, requires that, aside from the cultivated space needed to support household members, additional land has to be cultivated to produce fodder or be used pasture areas. *Cuemitl* may have taken longer to elaborate, restricting the amount of land that could be worked, but there was no need to feed or pasture animals. Instead, maize stalks probably were reincorporated into the field as a fertilization method, thus lowering the amount of additional natural fertilizer that needed to be applied.

In prehispanic times, neither traction animals, animal manure, nor the plow were available to indigenous populations. Major energy consuming tasks like turning up the soil —which today corresponds to the barbecho task— had to been done employing human labor. Many indigenous tasks have not been displaced by the introduction of the plow. The reason is that traditional techniques do not interfere with the newly incorporated technologies and thus they have been retained. For example, sowing maize in groups, as opposed to spreading them out like old world crops (wheat and rye), is clearly an indigenous development. Traditional maize landraces need to be grouped in order to avoid falling (*acame*) and provide plants with more stability. Moreover, intercropping methods are still practiced because the benefits of some plant associations are regarded as beneficial for soil fertility and for the production of various subsistence products.

Unfortunately, in some areas the patterns of indigenous land use are changing rapidly. Methods that where still prevalent 30 years ago like intercropping, plant transplanting and manual prolificacy suppression in maize are now becoming lost. These techniques made more efficient use of the available resources, and helped to mitigate local environmental constraints given the available technology. Yet, with the advent of modern practices, economic crisis, migration, and better job opportunities in urban cities for the younger members of the household, there has been a gradual abandonment of the countryside. Older people now work the agricultural fields and there is practically an absence of young adults. Knowledge of many of the techniques described in this chapter is not being transmitted to new generations and most of it is destined to disappear.

Chapter 6
Agricultural Production for the Year 2009: the Ethnographic Survey

This chapter deals with the ethnographic survey data on agricultural production for the year 2009. The focus of the survey was to record yield information from agricultural fields of 24 households within the study region in relation to quotidian human management strategies and differential climatic events. The methodology was designed as an observational study, which resembles an experimental one except that manipulation of the data occurs naturally rather than being imposed by the researcher (Utts 2005: 59).

Initially, the survey was designed to include 30 households from 10 different towns within a 540km² region. This area corresponded to the archaeological survey boundary established by the Acatzingo-Tepeaca Project (PAT) initiated by Dr. James Sheehy and Dr. Kenneth Hirth of Penn State University (Figure 28). Instead of representing a cultural or political area, it was intended to sample cultural development across all time periods in the Tepeaca region. While initiating the ethnographic work I realized that limiting my study to the area of the archaeological survey would leave out important environmental areas previously contained within the 16th century AD Tepeaca *altepetl* territory. I reasoned that employing the prehispanic territorial boundaries instead of the archaeological ones would provide a more informatively comparative picture of how production may have fluctuated across all of the altepetl's core territory. Therefore, the surface area was extended to 1300km² and it also included the southern portion of the Llanos de San Juan, and the eastern portion of the Valley Puebla-Tlaxcala (Figure 28). Obviously, this changed the structure of how data was collected. Not only did more area have to be considered, but also additional time and resource management had to be rearranged due to the larger traveling distances.

The goals of the survey were to obtain information on various aspects of traditional agricultural practices. This would serve to broaden our understanding and identify the

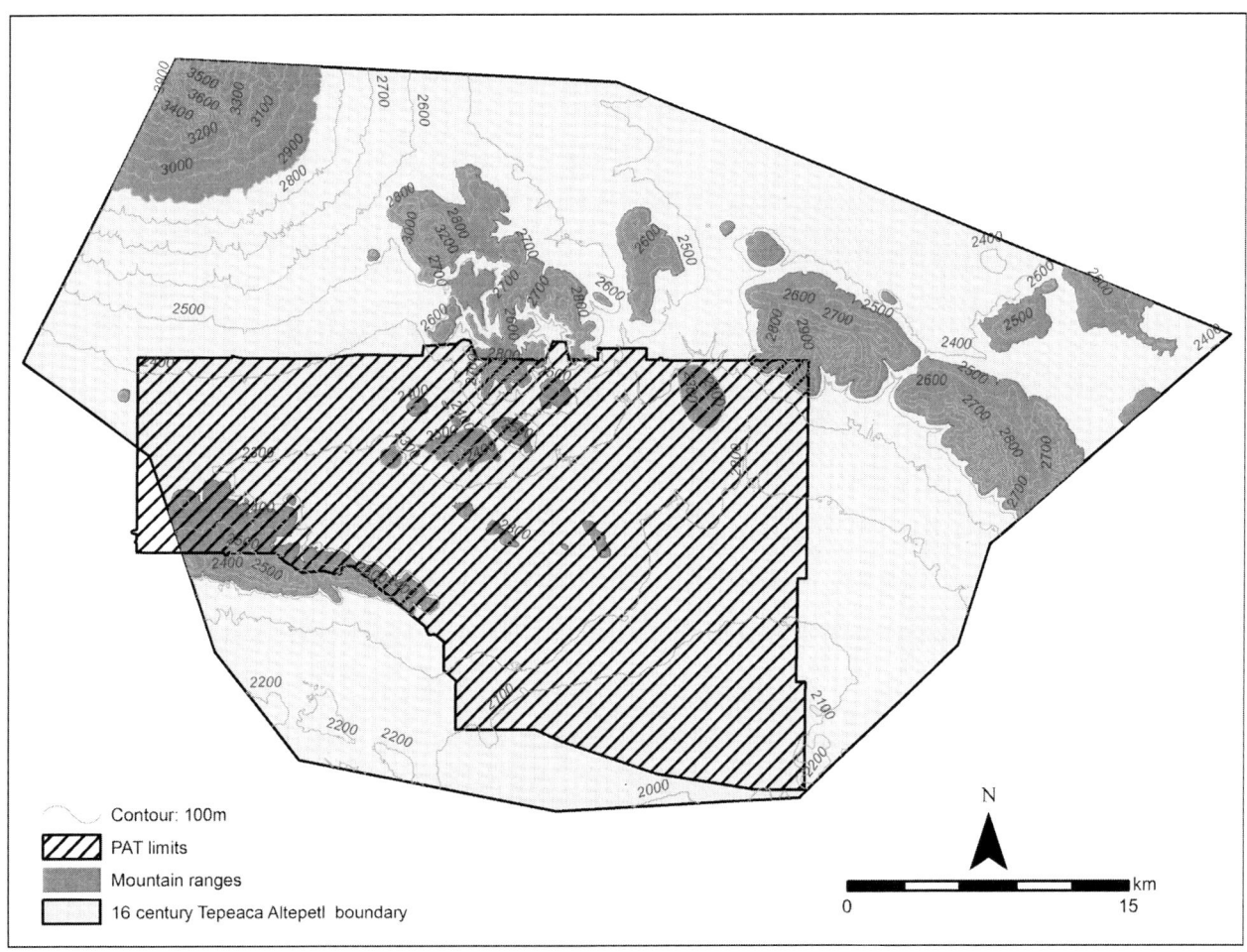

FIGURE 28. MAP SHOWING THE PAT PROJECT AND THE 16TH CENTURY AD TEPEACA ALTEPETL BOUNDARIES

key variables that influenced the differences in production between single households at the regional scale. Data were collected through personal interviews with local farmers from January to March of 2009. Information was gathered using both a survey questionnaire and informal conversations with farmers on topics such as types of crops used, fertilization methods, climatic events affecting the crops, the amount of work inputted, and total yields.

Harvest time, occurs from September to December and time was programmed for quantifying yields for the fields of each surveyed household. From March to September I had planned to make several visits to these fields and observe overall crop development in relation to the 2009 climatic regime and to record the negative effects of pests, disease, or lack of work. Unfortunately, in June the canícula struck with great intensity. The canícula is a system of high pressure that develops over the central Mexican Highlands and produces a decrease in rainfall that varies yearly (see Chapter 3). During the summer of 2009, this produced a severe mid-season drought and typical rain patterns were disturbed between one to three months depending on the area. No precipitation fell during the crucial period of ear development stages for maize and a large proportion of plants became barren or died out. After the lull, rains came back in such great quantities that they weakened the plant's root systems causing them to rot due to water over saturation in the subsoil. In areas with good drainage in alluvial and sandy soils, there was an outburst of weeds that resulted in high nutrient competition with the crops. Large numbers of farmers made no effort to control weeds and the result was that many crops stands were abandoned.

The fact that a high percentage of fields were abandoned had important repercussions for this study. The initial study of the production variation for a usual year turned out to be one focused on recording a severe seasonal climatic disruption with a widespread decrease in food yields. Of course, neither I nor local farmers had foreseen this problem, but I found this to be an excellent opportunity to evaluate the extent and severity of the canícula effect and whether it had struck equally within the entire region or only in certain zones. But measuring variation and drought severity was not an easy task considering the extent of the study region. Therefore, I reasoned that a second larger survey would be needed. In addition to the 51 fields recorded in the household survey, I also calculated maize production from a sample of an extra 449 fields located within the study region. The chore was done from September to December of 2009. In the pages that follow, I describe the methodology employed for the ethnographic study and the results obtained.

Part one: the household survey

Methodology

A first part of the study focused on measuring production from a sample of agricultural plots belonging to different households within the study region. It is imperative to emphasize that the objective was to quantify total yields in relation to different environmental settings and managing strategies. Therefore, the survey concentrated on extracting the overall configuration of the agricultural plots and their work inputs and not on household structure, its configuration or any other aspect not related to agriculture. Because a relatively large area was to be covered, the sampling design was configured to obtain the most information possible from different environmental settings. For this reason, I stratified the region (Kalton 1983) using modern Mexican political municipality boundaries named municipios. Each municipio is composed of a head town and a territory with several dependent towns. Modern political configurations have changed considerably among central Puebla populations since the Late Postclassic. Nonetheless, even though territorial boundaries have been reconfigured in size, extent, and number of units these units are the heirs of the indigenous settlements that were subjects of the ancient Tepeaca *altepetl*. This is especially true for those towns that have retained their Náhuatl name, such as Tepeaca, Tepatlaxco, Hueyotlipan, and Mazapiltepec.

There are 26 municipios with all or part of their boundaries within the ancient Tepeaca *altepetl*. Portions of two of them, Teolocholco and Huamantla, are located at the northwest corner of the original Tepeaca *altepetl* and lie at altitudes of 3000m or more, well above the limits for agricultural practices; for this reason, they were not considered for the survey. One household was surveyed from each of the remaining 24 municipios. I did not preselect the household or its particular fields since farmers visit their agricultural fields intermittently during the agricultural cycle (see Chapter 5). Instead, I reasoned that most efficient way to interview a peasant household was to find one in the field while they were carrying out their agricultural chores. In order to have certain randomness, I plotted the municipalities and then generated a random point within each one. From this point, I drew a 2km diameter circle (Figure 29) which then became the focus area for searching and meeting with a local farmer.

Visits to the selected areas took place from January to March of 2009. Farmers were approached while performing their agricultural duties and asked if they voluntarily wished to participate in the survey. All the information on the goals of the study was provided in the native Spanish language. No preference was given to age, genera, or the technology employed. The only requisites in order to participate in the survey were that household's practice rainfall agriculture and that they planted their fields with local maize varieties. Since I was not acquainted with how many plots each household would have, or their precise location, this provided additional randomness to selecting the sample.

Because this was an agricultural study focusing on crop production, no personal data of the participants was recorded. Detailed information about the research and its goals was provided to the people. Both the questionnaire and the informal conversations that occurred during

Agricultural Production for the Year 2009

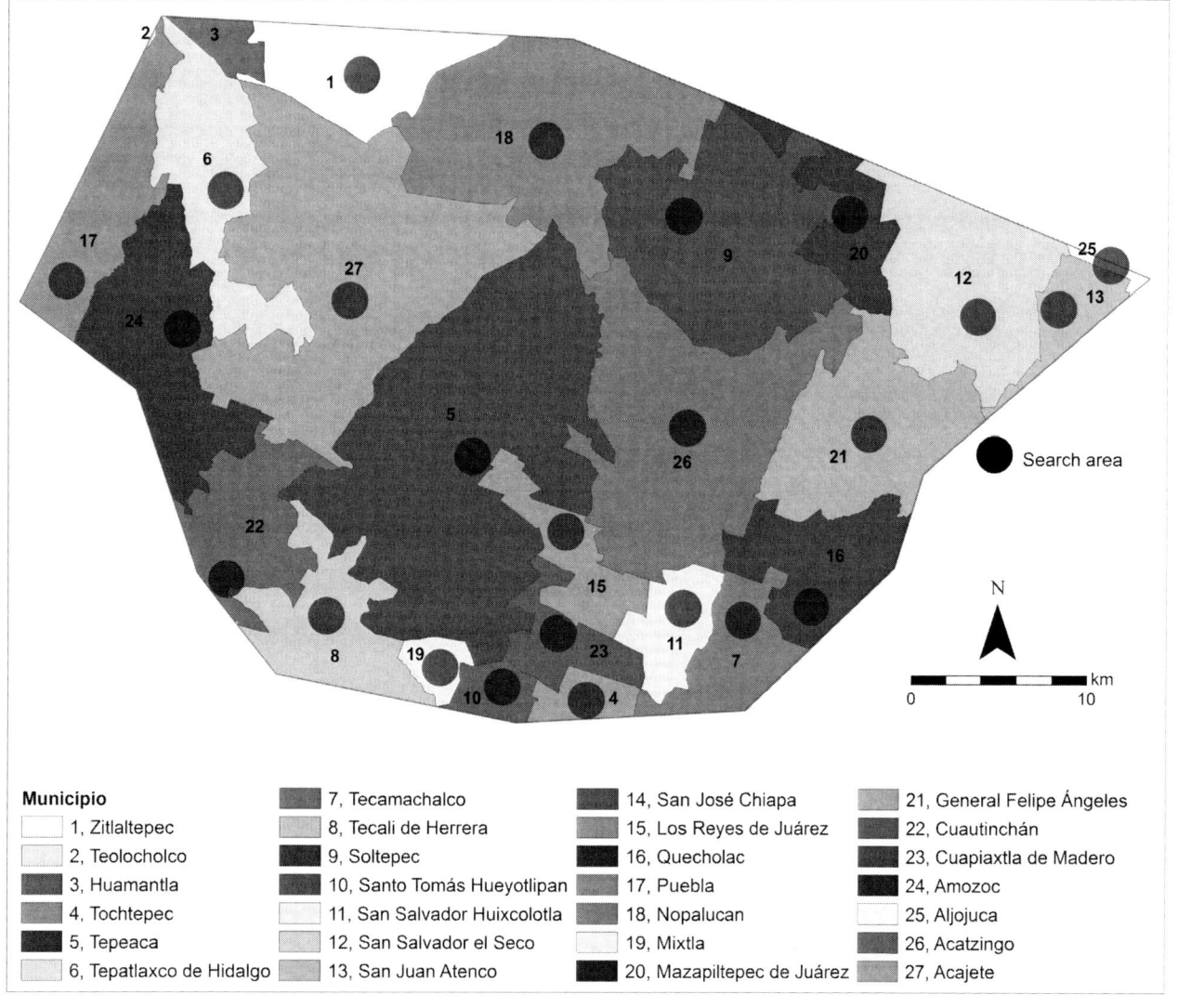

FIGURE 29. MAP OF THE STUDY REGION SHOWING MUNICIPIO BOUNDARIES AND THE SEARCH AREAS SELECTED THROUGH THE RANDOM POINT GENERATION.

the interview sessions centered only on agricultural information. To identify the fields of the different households I assigned a number for each one followed by the suffix 'Field' with its correspondent number (e.g., Household 1: Field 1 or Household 2: Field 2).

If the farmer agreed to participate in the study, then general information was recorded in relation to the locality, the field's number (out of the total), localization with GPS, the date and the field's area. Afterwards, the simple questionnaire consisting of five questions was completed. Question 1 asked what type of fertilizer was applied to the field and in what quantity? The farmer only had to answer what type of fertilizer he applied to the field -if any- and in what quantities.

Question 2 inquired how much rainfall precipitation had the field received during the 2009 year. Five possible answers were provided and these were 1) in excess, 2) plenty, 3) sufficient, 4) sparse, and 5) insufficient. This is a subjective measurement and depended on the farmer's perception of what is the optimal amount of water his/her field needs for good crop development. However, the information is relevant because it allows comparisons between the perception of the farmer and the soil composition of the lands that are used.

Question 3 was used to obtained data on negative effects on the maize stands. The question asked if the field was affected by: 1) frost, 2) pests, 3) hail, 4) theft or other variables which they had to explain.

Question 4 asked if the amount of work invested in the field was considered to have been sufficient or insufficient. They had four options from which to choose: 1) a lot, 2) sufficient, 3) low and 4) very low. This question sought to obtain information about the perception of how much work is needed for success in relation to idiosyncratic views of agricultural labor.

Question 5 asked to quantify the yields of a particular field during the previous 2008 year or further back if possible. Farmers were requested to give a rounded figure in kilograms and, if it was the case, all the products that were cultivated in the plot.

When present, additional fields were visited, located on a map, and their general features were recorded. Additional data, such as the plot extension, climate type, soil classification, and precipitation were extracted from previous studies and digital data databases (see Chapter 3 for details on the environment settings).

Measuring maize production

A core facet of this research was to register maize productivity for each of the fields surveyed. This was not a simple task due to the large area that had to be covered. Also, time management issues where problematic, in particular when visiting the fields at harvest time. Farmers can estimate the date when harvest will occur, but do not adhere to a strict date to perform it. It is largely determined by the available labor, and their continuous observations of climatic behavior. It also depends on the farmer's day-to-day evaluation of when maize ears will arrive at the right stage of dryness for harvest. Visiting two or three fields may not be much of a problem, but because I had to visit more than 50, inevitably harvest time for some households would overlap. An option was to let farmers harvest without me being present, and subsequently to visit them and weigh their total production back at the household facilities. However, this posed other problems because I would have needed to wait an interval of time for ears to have dried completely in order to weight them. This was problematic because, it was possible that by the time I visited the farmer's house some amount of maize would have been processed, given away, consumed, or sold. I would not have been able to determine how much was missing due to any of these factors, thus forcing my calculations to be inaccurate.

For this reason, I had to develop an efficient method that could permit me to do relatively fast in-field estimations of production yields with a reasonable degree of precision. In her famous study in the Oaxaca Valley, Ann Kirkby (1973) showed that there exists an important relationship between the length of a maize cob and its equivalent in grain weight. Although the study focused solely on maize from that region, which is environmentally different in some respects to the Tepeaca Region, I reasoned that such a relationship might also exist for Tepeaca. If it was possible to determine a regression equation for this relationship, and if plant producing densities per hectare were computed in the field, then it was possible to estimate yields using only a sample of maize yields for each plot any time after maize ears had reached maturity.

I began to develop this measure during the course of 2009 by collecting ears (from the 2008 harvest) of different varieties from several areas and measuring them. The goal was to find a significant relationship between ear dimensions and its equivalent in weight of dry shelled maize. In total, I obtained and measured 170 ears. As I point out below, cob length is a useful measurement, but two factors that are much more efficient indicators of dry grain production are ear volume and ear length. In particular, ear volume exhibits the strongest relationship with total weight of shelled maize, and thus became the basis for calculating maize production per hectare.

I wish to emphasize that neither volume nor length of the ear can by itself determine final maize production per hectare. In other words, we cannot effectively estimate how much a field produced based solely on the average dimensions of an ear or the cob. Plant densities per hectare, or more accurately the number of plants that were able to produce an ear, need to be considered if a precise estimate is to be obtained.

In the following section, I describe the methodology for ear length and ear volume calculations and some of the problems that I encountered when determining productivity based solely on weight. I then propose a method for estimating maize yields in the field, which combines ear volume, ear weight, and plant densities, to generate useful comparative figures for estimating maize productivity under conditions of rainfall agriculture. Since the method is based strongly on calculating the volume of the ear, I call it the Ear Volume Method (EVM).

Problems using weight measurement

Determining maize based solely on whole cob weight can be problematic due to the amount of humidity in both kernels and cobs. This is a key variable that must be considered when performing any type of yield calculation. Water content in kernels will depend on several factors including its developmental or maturation stage, the number of days it has been left to dry, the amount of precipitation in any particular year, or even fluctuating atmospheric humidity levels.

Shelled maize can be stored and preserved efficiently at its driest stage, when kernels have 14% humidity content. This is the level at which embryos can survive for long periods and the endosperm preserved. In addition, it is the standard percentage employed in agronomy and in economic commercial transactions when weighting fully dried maize.

Weight is a useful measurement for maize yield comparisons both locally and regionally. Unfortunately, the unstable nature of maize moisture content during the two to four months after it has arrived matured may produce a considerable amount of measurement error. Technically, if weight is determined from ears or shelled maize, then optimally it should be taken when it is at an average of 14% humidity (Instituto Nacional de Investigaciones Forestales 1997).

Farmers know when it is the right time to harvest dry maize ears. A key indicator is when the ears are on the maize stalk and they start to bend downwards with the tip pointing to the ground. When this occurs, maize ears are fully mature and ready for picking. By this time, moisture content may be around 16-19% or lower. In addition, when trying to shell maize, if the kernels do not detach easily, they know there is still too much moisture in the ear and they must be left to dry for a longer period. Unfortunately, it may take a long time for ears to arrive at the desired 14% water content. It largely depends on climatic factors such as sunlight and radiation, or even the time of the day when the ears are collected. Farmers know that ears of maize will weigh more if they are measured in the morning hours (before 10 am) or on a cloudy day compared to the afternoon hours on a sunny day when solar radiation is at its strongest creating high rates of evaporation. Light scattered showers, which can occur any time after the harvest period can raise moisture content to as much as 25% or more. These problems have to be taken into consideration if maize is to be accurately weighted in the field.

To obtain a valid measurement for maize production requires weighing maize at the right moment in time, but determining when this moment is may take considerable time to determine. If the researcher needs to calculate yields form several plots dispersed over wide areas, time management issues become a problem. For these reasons, I found the method for calculating weight of shelled maize based on the volume of the ear especially useful for the reasons that I explain below.

Volume as an alternative option for determining production

The method I developed for estimating maize production is based on the volume of the ear and the number of plants that produced an ear. I will first describe how to calculate the volume of the ear and its equivalent weight of dry kernels. The volume of the ear can be determined using the formula for a truncated cone:

$$V = \pi h/3 (R^2 + r^2 + Rr)$$

Where:

V = volume
$\pi (pi)$ = 3.1416
R = Radius of the base
R = radius of the top
H = height of the truncated cone

The measurements need to be taken in the following way (Figure 30). The radius of the base has to be obtained at the point where it curves inwards towards the bud (the lower end of the ear). This is usually the middle point of the third or fourth line of kernels from the bottom. The radius of the top is also calculated where the kernels curve inwards towards the tip, which is generally at the middle point of the second or third line of kernels from the tip of the ear.

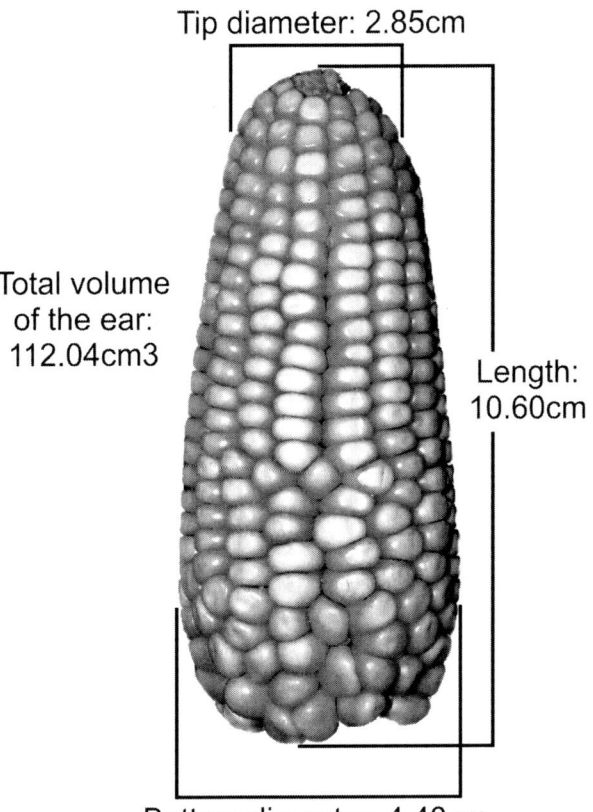

FIGURE 30. MEASUREMENTS TAKEN FOR DETERMINING THE VOLUME OF AN EAR.

The length is measured from the bottom part of the first line of kernels at the base of the ear, to the top portion of the first line of kernels at the tip. Note that for estimating the volume of the ear the length measure does not correspond to the cob length. The reason is because many times the upper portion of the cob surpasses the first line of kernels and thus it will give an unbalanced ratio between volume of the ear and the weight of kernels (e.g., Figure 31: F).

As mentioned before, I measured 170 ears of maize that were collected throughout the study region. Although I intended to collect a larger sample, I found that the correlation between volume and weight of shelled maize was so strong that measuring more ears soon proved to be redundant. The ears varied considerably in shape, size, row number, and length (Figure 31). This provided good variance of dimensions on which to calculate volume and weight. Ears with well-developed kernels were preferred, but several cobs exhibited anomalies, were underdeveloped, lacked kernels or had pest infestation such as *palomillas* (e.g., *Helicoverpa zea*). The ears had been harvested during the 2008 agricultural cycle and stored for several months. Thus, their moisture content averaged about 14%. To ensure constant moisture content, all cobs were put in the sun and left there for two to three days. They were weighted before and after sun drying and it was corroborated that their weight totals only varied between 0.1 and 0.3 grams between both measurements, which is exactly the kind of variance that is registered

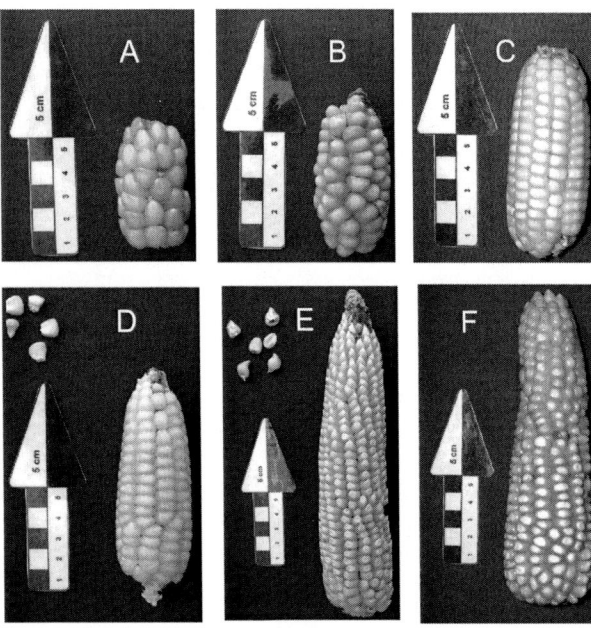

FIGURE 31. EXAMPLES OF MAIZE EARS FROM THE STUDIED REGION.

when the kernels cannot be dried any further (Instituto Nacional de Investigaciones Forestales 1997).

Subsequently, both a bivariate correlation and a regression analysis were performed with SPSS software (Statistical Package for the Social Sciences). The Pearson Correlation (2-sided) test was calculated for Total Volume of the ear and Weight of the Kernels (without the cob) and produced values of $R=0.997$ and $R^2=0.993$ (Figures 32-34) and significant values for both calculations were at the 0.01 level. The regression analyzes indicates that the constant value or intercept for Weight of the Kernels was -0.35 and the coefficient for Total Volume was 0.833.

The formula obtained from the sample for determining the weight of dry kernels based on the volume of the ear is:

Weight of dry kernels = -0.03491 + (0.8333 x volume of the ear)

Correlations		Volume	Weight Kernels
Volume	Pearson Correlation	1	.997**
	Sig. (2-tailed)		.000
	N	170	170
Weight Kernels	Pearson Correlation	.997**	1
	Sig. (2-tailed)	.000	
	N	170	170

FIGURE 32. RESULTS OF THE CORRELATION ANALYZES RUN IN SPSS SOFTWARE.

Roughly speaking, what this means is that an ear with a volume of $100 cm^3$ will produce 83.33 grams of shelled maize at 14% humidity content.

Length of the ear and weight of the dry kernel

On the other hand, Length of the ear and Weight of the Kernels also has a good correlation of $R=0.95$ and $R^2=0.90$ at the 0.01 significant level (Figures 35 and 36). The line fit is not as good as the one for Total Volume and Weight of the Kernels, but there is still a very good correlation.

Length of the cob and weight of the dry kernel

Length of the cob and its conversion to weight was the relationship that Ann Kirkby (1973) used for determining ancient maize production capacity for the Valley of Oaxaca. For the Tepeaca region, Length of the cob in relation to the weight of dry kernels also has a good correlation of $R=0.926$ and $R^2=0.858$ significant at the 0.01 level (Figures 37 and 38). However, determining the weight of shelled maize based solely on the length of the cob poses a problem. While Length of the Ear is the measurement of the maximum extent of kernels, Length of the Cob is the total length only of the cob. This is an important diffcrence because commonly the upper and middle area of the cob lack kernels due to several factors such as nutrient and water deficit, or as the result of kernel loss due to pest infestation (see Figure 31: F). The high correlation between cob length and shelled maize that I determined with the regression analysis only works if most of the cob has kernels. Thus, measuring the length of the cob does not guarantee that an adequate estimation of shelled maize can be obtained.

From the results of the regression and correlation analyzes of the 170 ears measured, I decided that the best option for determining yields from the agricultural fields surveyed was to use the value obtained from measuring the volume of ears of maize. Using this measurement has several benefits in relation to weighting the amount of shelled maize produced per area. First, it is time efficient because only three simple measures need to be taken from ears. Second, the measures can be obtained before the ears are harvested. The only requisite is that the procedure is made after the 185 day maturity period for six month maize varieties, and 130 day period for the four month varieties. After this period, the ears are fully developed and there essentially will be no changes in maize cob dimensions. Third and finally, there is no need to worry about moisture content because volume can later be transformed into weight using the regression analyzes discussed above.

Calculation of maize production per hectare

Once I had established the relationship between volume of the ear and its equivalent in weight of shelled maize, I used these values for calculating yields per hectare for each of the fields of the 24 households. The methodology employed for estimating maize production was derived

Model Summary and Parameter Estimates							
Dependent Variable: Weight Kernels							
Equation	Model Summary					Parameter Estimates	
	R Square	F	df1	df2	Sig.	Constant	b1
Linear	.993	23979.97	1	168	.000	-.035	.833
The independent variable is Volume.							

FIGURE 33. RESULTS OF THE REGRESSION ANALYZES RUN IN SPSS SOFTWARE FOR TOTAL VOLUME OF THE EAR AND WEIGHT OF THE KERNELS.

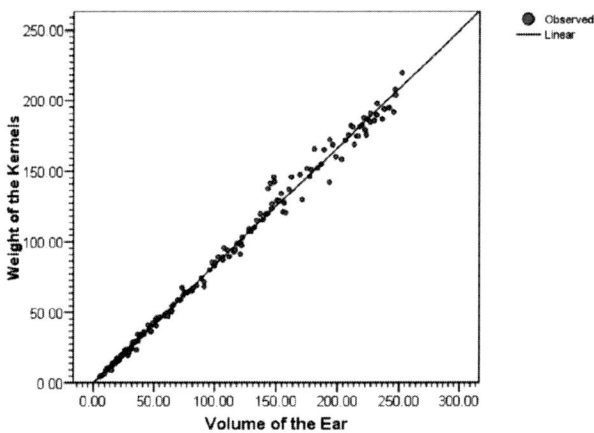

FIGURE 34. REGRESSION ANALYZES FOR TOTAL VOLUME OF THE EAR AND ITS EQUIVALENT WEIGHT OF THE DRY KERNELS.

Model Summary and Parameter Estimates							
Dependent Variable: Weight Kernels							
Equation	Model Summary					Parameter Estimates	
	R Square	F	df1	df2	Sig.	Constant	b1
Linear	.905	1607.913	1	168	.000	-50.564	14.319
The independent variable is Ear Lenght.							

FIGURE 35. RESULTS OF THE REGRESSION ANALYZES RUN IN SPSS SOFTWARE FOR LENGTH OF THE EAR AND WEIGHT OF THE KERNELS.

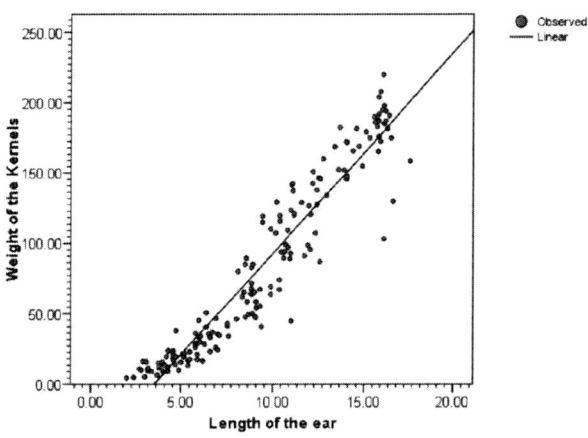

FIGURE 36. REGRESSION ANALYZES FOR LENGTH OF THE EAR AND WEIGHT OF THE DRY KERNELS.

Model Summary and Parameter Estimates							
Dependent Variable: Weight Kernels							
Equation	Model Summary					Parameter Estimates	
	R Square	F	df1	df2	Sig.	Constant	b1
Linear	.858	103.345	1	168	.000	-46.855	13.053
The independent variable is Cob Length							

FIGURE 37. RESULTS OF THE REGRESSION ANALYZES RUN IN SPSS SOFTWARE FOR LENGTH OF THE COB AND WEIGHT OF THE KERNELS.

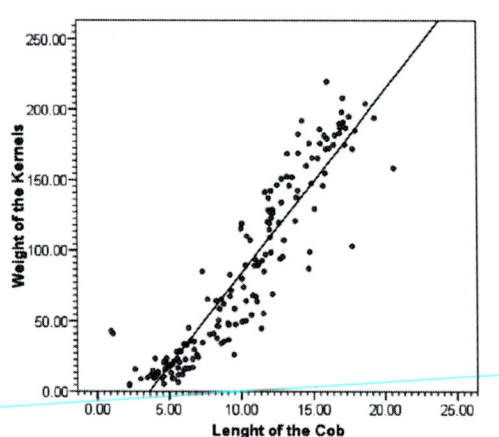

FIGURE 38. REGRESSION ANALYZES FOR LENGTH OF THE COB AND WEIGHT OF THE DRY KERNELS.

in part from agronomic survey techniques. For each field, two tracks of 10m were walked, following furrow orientation -which is usually the longest portion of the field. On each track were counted the number of plants that produced an ear and the number of barren plants. This provided information on the ratio between the number of surviving plants and the number of producing plants. Each ear encountered was unhusked by puncturing the tip of the ear with a large needle that locals use when harvesting, and then pulling down a portion of the husk. For each ear, its tip, base and length were measured. In addition, the number of plants within each transect were counted and their average separation.

Therefore, calculation of production for a field was made by:

$$P = AVSME \times PD$$

Where:

P = Production
$AVSME$ = Average Volume of Shelled Maize per Ear
PD = Plant Density

For the Average Volume of Shelled Maize per Ear (AVSME) value, I used the average volume of the ear per each particular field. Total weight was also calculated based on this measure.

Plant density was determined by the following:

$$PD = CA \times NMP/SA$$

Where:

PD = Plant Density
CA = (constant area of $10,000m^2$)
NMP = Number of Maize Plants (per sampled area)
SA = Sampled Area

In which:

$$SA = FW \times LST$$

Where:

SA = Sampled Area
FW = Furrow Width
LST = Length of the Sample Transect

An additional ratio value was obtained by dividing the number of plants that produced an ear by the total surviving plant density per hectare. Although it may not be an exact reflection of the season's climatic regime and its effect on cultivars, this ratio supplies an idea of the percentage of plants that were affected by environmental factors, pests, theft, or other conditions, as in the case of the 2009 canícula drought.

The Ear Plant Producing Density Percentage (EPD%) is the number of plants that were able to generate an ear in relation to plant density estimation. It was calculated by:

$$EPD\% = NPE/PD$$

Where:

$EPD\%$ = Ear Producing Density %
NPE = Number of Plants with an Ear
PD = Plant Density

Initial plant densities, survival plant densities and total productivity

An initial Plant Density (IPD) was calculated for each field. This corresponds to the estimated number of kernels that were planted per hectare. The estimation was based on an average deposition of three kernels per plant group, which is the average number used today by farmers in the Tepeaca Region. Thus the Initial Plant Density is determined by:

$$IPD = NKGP \times NPGh$$

Where:

IPD = Initial Plant Density
NKGP = Number of Kernels per Group Plant
NPGh = Number of Plant Groups per hectare

In which:

$$NPGh = NPGFL \times NFPh$$

Where:

NPGFL = Number of Plant Groups per Furrow Length
NFPh = Number of Furrows per hectare

The Plant Surviving Density Percentage (PSD%), or those plants that germinated and developed from the initial planting, was calculated by:

$$PSD\% = PD / IPD$$

Where:

PSD% = Plant Survival Density percentage
PD = Plant Density
IPD = Initial Plant Density

Result

A total of 51 fields from 24 households were surveyed (Figure 39). Each household had between one to three rainfed dependent fields scattered throughout the landscape. The mean number of fields per household was 2.13 (this number does not include the presence of a house lot cultivation plot).

The area of the fields cultivated by the 24 households ranged from 0.21ha to 3.44ha in size (Figure 40). Though the mean amount of land cultivated per household was 1.54ha, there were some outliers. Household 2 from Amozoc had the highest amount of land, totaling 3 fields that measured 3.44ha. Households 19 from Zahuatlan and 21 from Acajete also had 3 fields that amounted to 2.53ha and 2.34ha respectively. On the lower side of the spectrum,

HH	Locality	Municipio	Number of fields	Total area cultivated (ha)	Total maize production (kg)	Average maize production per field (kg)	Average yields from all fields (kg/ha)
1	Capulac	Puebla	2	1.95	876	438	421
2	San Mateo	Amozoc	3	3.44	1838	613	523
3	Almoloyan	Cuauhtinchan	2	1.69	164	82	97
4	Ajajalpan	Tecali	2	1.75	169	85	96
5	Mixtla	Mixtla	2	0.67	46	23	73
6	Hueyotlipan	Hueyotlipan	1	0.21	9	9	42
7	Caltenco	Tochtepec	2	1.35	121	60	75
8	Tlaixpan	Tecamachalco	3	1.26	104	35	89
9	Las Tunas	Quecholac	2	1.28	120	60	93
10	Santa Ana	El Seco	2	1.63	693	346	426
11	Atenco	Atenco	2	1.52	787	394	533
12	Tecuitlapa	Aljojuca	1	0.63	290	290	462
13	Xicotenco	Mazapiltepec	1	0.45	448	448	995
14	Obregon	Soltepec	3	2.23	1458	486	699
15	Ixtiyucan	Nopalucan	2	1.03	834	417	852
16	Tepulco	Tepatlaxco	2	1.46	825	413	494
17	Villanueva	Acatzingo	2	1.64	142	71	85
18	Acozac	Los Reyes	2	1.01	129	65	130
19	Zahuatlan	Tepeaca	3	2.53	676	225	267
20	Huixcolotla	Huixcolotla	2	1.38	105	52	88
21	Acajete	Acajete	3	2.34	1079	360	439
22	Citlaltepec	Citlaltepec	2	1.37	866	433	604
23	Tenango	Felipe Angeles	3	2.43	517	172	204
24	Cuapiaxtla	Cuapiaxtla	2	1.67	110	55	67
Average			2.13	1.54	517	235	327

FIGURE 39. FIELDS MEASURED FROM THE ETHNOGRAPHIC SURVEY OF 24 HOUSEHOLDS.

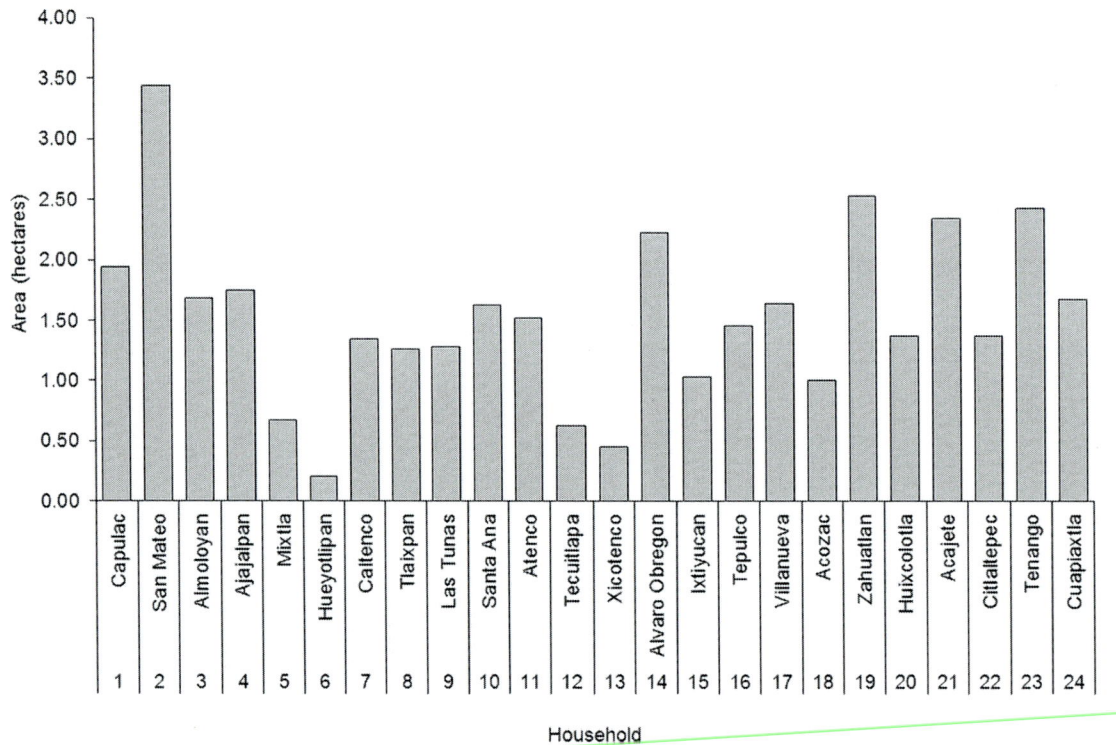

FIGURE 40. TOTAL AREA CULTIVATED BY HOUSEHOLD.

Household 6 from Hueyotlipan had only one field (0.21ha) and it was the smallest plot in the sample. Two more had one field, Household 12 from Tecuitlapa and Household 13 from Xicotenco. In both these cases, I was told that even though they possessed relatively little land, they kept working plots because their cultural tradition demanded it and not because it provided the principal source of food for them. Working the plots was considered an extra source of income because the households were engaged in other commercial activities that comprised their main source of sustenance.

Plot shapes where mostly rectangular, but triangular and irregular forms were also present. Their size was relatively small ranging between 0.21ha to 1.36ha with a mean of 0.71ha. Although I cannot explain the reasons for the relatively small sizes of the fields, I suspect that continuous population growth has led to land scarcity and progressive subdivision of available lands. In particular fragmentation and the concomitant shrinkage of field size has tended to increase in recent years throughout central Puebla. Fields that once comprised large areas and belonged to one household often are now subdivided between several heirs. Consequently, today's landscape is crowded with numerous small plots. Many are of 1/2 to 1/8ha in size. In some areas, parcels are even smaller, resulting in farmers having only four to eight lines of furrows in each. Fragmentation has been ameliorated somewhat by land selling, especially by persons that have small tracks of land. Generally the first option is to offer to sell the land to a sibling or some other kinsmen so as to maintain the largest amount of land in a family. A second option would be to sell land to a friend in the locality, particularly if they can negotiate a good price for the property. Otherwise, it is sold to anyone interested in buying land. However, transactions generally stay in the community because the prevalent ejido land tenure system prevents, in theory, the privatization of lands.

Production from the surveyed field

One aspect that is striking about central Puebla and Tlaxcala is that most land is cultivated on a year-by-year basis. There are very few spots where land is under fallow, and usually for no more than one year. This almost certainly affects the production capacities of fields, especially for local rainfed agricultural systems. Total production per household for 2009 ranged from as low as 9kg for Household 6 from Hueyotlipan, to 1838kg/ha for Household 2 from Amozoc (Figures 41 and 42). Also, Households 14 (Obregon) and 21 (Acajete) produced above one metric ton of shelled maize. Even if total production for Household 2 (Amozoc) was good and well above the nutrition requirements it required for one year, its average productivity was low reaching only 523kg/ha. On the contrary, Household 13 from Xicotenco had an average production of 995kg/ha, which is much better. However, it only cultivated 0.45ha and therefore its total production for the year was low (448kg total).

Soils and production

In some instances, having more than one field aids to level out productivity. Not only does this permit coping with

Agricultural Production for the Year 2009

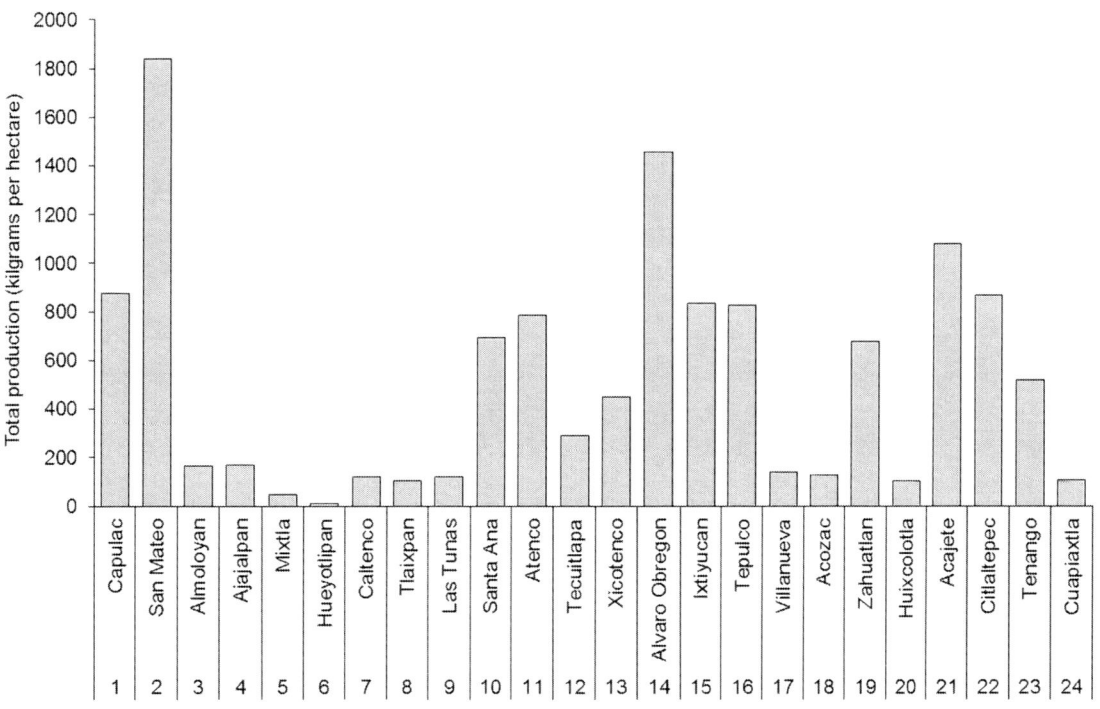

FIGURE 41. TOTAL PRODUCTION PER HOUSEHOLD. IT INCLUDES THE PRODUCTION FROM ALL THE FIELDS UNDER CULTIVATION.

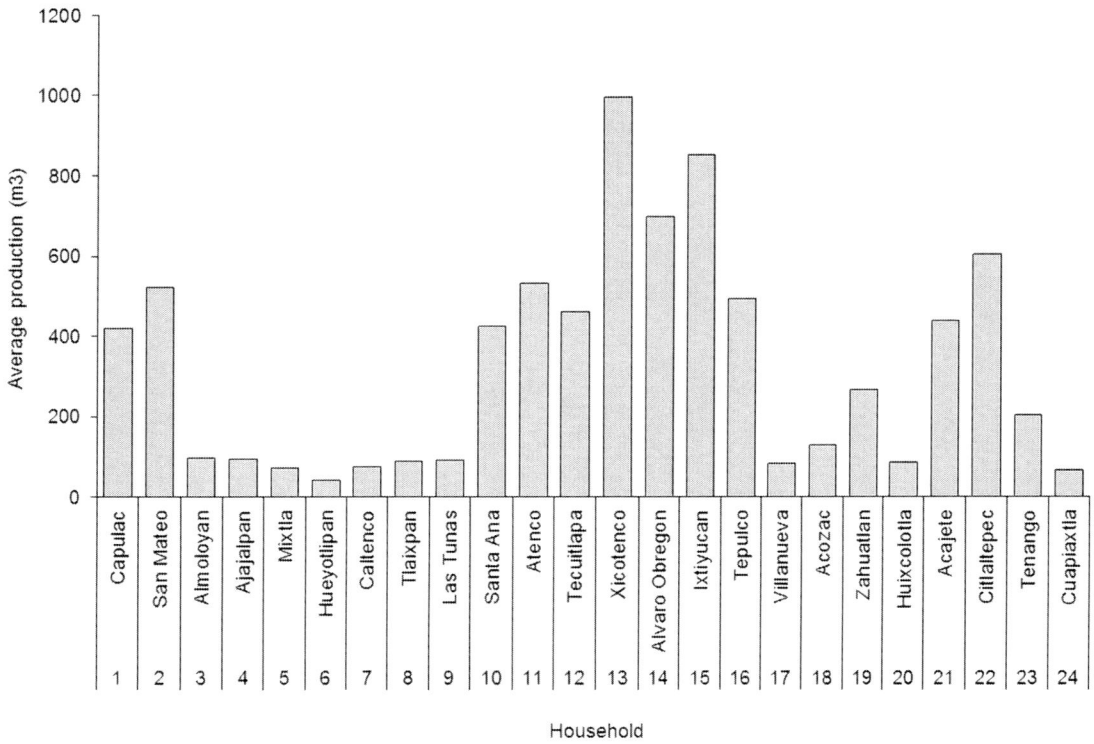

FIGURE 42. AVERAGE PRODUCTION OF MAIZE PER HECTARE FOR EACH HOUSEHOLD.

differential climatic hazards such as frosts and hail, but also it helps to offset variable rain patterns in time and space. Having fields in different areas also allows cultivating in different soil types. This is especially important when there are contrasting soils qualities in terms of texture and workability. In order to evaluate soils in relation to total productivity by field, I developed a scale value using soil distribution in the studied region and the descriptions and characteristics of each type (see Chapter 3). For each soil, a scale value was assigned that ranged from 1 to 5, 1 being the most useful, workable, and deepest soil type, and 5 the poorest and shallowest. According to regional soil

Soil key	Primary soil	Secondary soil	Tertiary soil	Physical phase	Average scale value
Hh/2	5				5.0
Be+Je+Hc/1	3	4	5		4.0
Be+Hh/1	3	5			4.0
Hh+Be/2	5	3			4.0
Vc/3	4				4.0
Je+Bv/1	4	3			3.5
Be+Bk+Vc/2	3	3	4		3.3
Be+Hh+To/2	3	5	2		3.3
Je+Lo/1	4	2			3.0
I+E+Be/2	1	4	3		2.7
I+Rc+E/2	1	3	4		2.7
Be+To+Th/2	3	2			2.5
I+E/2	1	4			2.5
Hh+Be+I/2/D	5	3	1	-1	2.0
Re+I+Hh/2/L	3	1	5	-1	2.0
Xh+Hc/2/PC	2	5		-1	2.0
Xh/2	2				2.0
Be+I+E/2/L	3	1	4	-1	1.8
Be+I+E/2/L	3	1	4	-1	1.8
Rc+Bk/2/L	3	3		-1	1.7
Th+Rd/1/P	2	3		-1	1.3
Th+To/2/L	2	2		-1	1.0
Th+To/2/L	2	2		-1	1.0
I/2	1				1.0
Xh+I/2/PC	2	1		-1	0.7
Xh/2/PC	2			-1	0.5

Soil Nomenclature				Texture		Physic phase	
Be	Eutric Cambisol	Lo	Orthic Luvisol	1	Coarse	P	Rocky
Bk	Calcic Cambisol	Rc	Calcaric Regosol	2	Medium	L	Lithic
Bv	Vertic Cambisol	Rd	Dystric Regosol	3	Fine	D	Duric
E	Rendzina	Re	Eutric Regosol			PC	Petrocalcic
Hc	Calcaric Phaeozem	Th	Humic Andosol				
Hh	Haplic Phaeozem	To	Ochric Andosol				
I	Lithosol	Vc	Chromic Vertisol				
Je	Eutric Fluvisol	Xh	Haplic Xerosol				

FIGURE 43. SCALE VALUES OF THE DIFFERENT SOIL TYPES PRESENT IN THE STUDY REGION.

maps from Mexico's INEGI, some fields had up to three types of soils. In addition, some areas have Rocky, Lithic, Duric, or Petrocalcic horizons not suitable for agriculture, in which case if present they were given a -1 value. By averaging the values of the primary, secondary and tertiary types of soils present in each area, I established an ordinal average scale value for soil types within the study region (Figure 43). This scale is useful in order to compare the potential of each zone for agriculture productivity.

Another variable that had to be considered was the fertilization regime employed. Fertilization methods varied within the study region according to their availability. The most preferred method was manure fertilization (cow, pig, chicken). Chemical fertilization is a very different strategy that is managed in its own terms. It does not need to be mixed during the barbecho or prior to sowing. It is usually applied prior to the ear development to provide the plant with sufficient nutrients for ear development. As explained in Chapter 5, it is considered too 'hot' and thus has to be applied under good moisture conditions in order to avoid burning plants. Yet, some people did not fertilize at all due to high prices on both types of fertilizers, particularly in fields with humidity retention problems or poor soil development. Interestingly, even though the potential yield of these fields was markedly low, around 400kg during an average year, farmers kept cultivating them. They usually stated that 'it is better to have something instead of nothing', and given that the seed came from a personal stock and monetary resources were expended, then even a small production of 300 to 400kg/ha make it worth cultivating on poor soils.

Generally, all fields from the same household were either fertilized in the same way or not fertilized at all. As I pointed out in Chapter 5, in some areas farmers have opted not to fertilize due to the high fertilizer prices. This is especially true when plots have poor soils. I have seen that, when a farmer has several fields with different types of soils, he tends to put more fertilization resources in the ones with the best conditions for cultivation (e.g., irrigation vs. rainfed, alluvial plain vs. piedmont). However, chemical fertilizers are generally set aside for fertilizing irrigated fields with hybrid maize varieties that are expected to yield higher and compensate for the expenses.

Therefore, if productivity is correlated with soil quality it would be expected that higher levels of production are obtained in the best soil types. Apparently for the 2009 year this was not so. A correlation and regression analyzes show that there is basically no correlation at all (Figure 44). Results gave R=0.17 and R^2=0.31. For example, fields from Household 21 where fertilized equally with cow manure but the yields differed considerably between them (Figure 45). In addition, even though some of the fields from this household had good soils (scale value of 4.00), they produced far less than fields that had soil of less quality (scale value 3.00).

The production from fields of Household 1 was markedly different. Both its two fields had soils with a scale value of 3.50. Yet, Field 2 was more than twice as productive much as Field 1.

The results of the correlation analysis between soil quality and maize yields show that the variance could not be explained by differential use of fertilization methods or by differential soil qualities. The explanation is probably, as I show below, in the differential rain patterns brought about by the canícula event in 2009.

Rain patterns and the 2009 canícula drought

In 2009, the low and disparate levels of production recorded for most of the fields of the 24 households surveyed were strongly influenced by the atypical rain patterns brought about the mid-summer canícula event. At the beginning of the season, good rains came and anticipated a successful agricultural year. Sowing tasks initiated in the western part of the study region between April and May giving farmers a good head start on planting tasks. This was important because it allowed the plants a longer time period to develop and a better chance of avoiding early fall frosts.

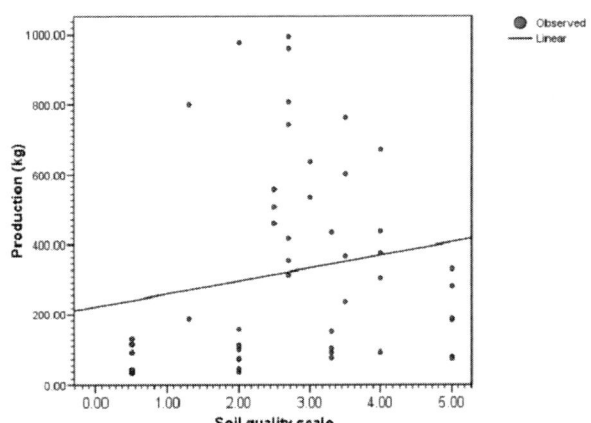

FIGURE 44. REGRESSION ANALYZES FOR PRODUCTION PER HECTARE AND SOIL QUALITY.

Household	Field	Soil type	Scale value	Maize yields (kg/ha)
1	1	Je+Bv/1	3.50	238
	2	Je+Bv/1	3.50	603
2	1	Be+Je+Hc/1	4.00	438
	2	Je+Bv/1	3.50	367
	3	Je+Bv/1	3.50	764
8	1	Xh/2	2.00	73
	2	Xh/2	2.00	35
	3	Xh/2	2.00	159
15	1	I+E+Be/2	2.67	960
	2	I+E+Be/2	2.67	743
21	1	Je+Lo/1	3.00	636
	2	Be+Je+Hc/1	4.00	377
	3	Be+Je+Hc/1	4.00	305
24	1	Xh/2/PC	0.50	42
	2	Xh/2/PC	0.50	92
Average				389

FIGURE 45. MAIZE YIELDS PER HECTARE FROM THE FIELDS OF SIX HOUSEHOLDS IN RELATION TO SOIL TYPE.

Farmers understand that around June-July the canícula enters the region. When the canícula arrives it comes as a dry-windy period with an overall decrease in rainfall. In El Niño years droughts can intensify especially due to an important decrease in humidity conditions and cloud formation over the central Mexican plateau, and an increase in the amount of radiation that the surface receives; on the contrary, La Niña years produce rain patterns near the average (Magaña *et al.* 2004). Unfortunately for farmers, this year the canícula coincided with the arrival of an El Niño event and the intraestival period came as a dry one. The canícula phenomenon is expected to produce a decrease in rains that usually lasts between two to four weeks. After this period rains return and normalize. Farmers estimate that the canícula will affect crops stands in some way, but its impact is usually minor because it strikes before the crucial ear development in maize and in inflorescence development in bean plants. Much to everyone's surprise, the 2009 canícula period continued for more than four weeks and the drought continued, probably intensified by the El Niño Southern Oscillation effect (ENSO) as has been observed in past times (Magaña et al. 2004: 56-57), and extending far beyond what maize plants can tolerate. INIFAP´s (www.inifap.gob.mx) data indicates that, in some areas, the drought persisted for more than two months (Figures 46 and 47). In the Tecamachalco vicinity, the last considerable precipitation (22.2mm) fell on June 28. From there on, only 16 minor showers occurred (averaging 0.4mm). It was not until September 4th, 69 days after the last considerable rainfall that a good shower occurred (14.4mm). Rains leveled out in September, but it was too late for the majority of plants that by this time had lost their ability to pollinate and would grow short and barren.

Perhaps even more surprising was that in some areas, when the rains returned they arrived as extremely hard downpours. The contrast was so sharp that in areas around Acatzingo it rained five times more in September than in August. The Tepeaca weather station indicates that the mean precipitation for 2009 was 654mm, a very similar amount to that of the previous 2008-year and well above the 2005 and 2006 seasons (Figure 48). A closer look at the data shows that in July and August of 2009 precipitation felt drastically as opposed to the four previous years. Thus, precipitation records exemplified that the problem for crops was not the total amount of precipitation during the year, but rather the time when it fell during the growing season.

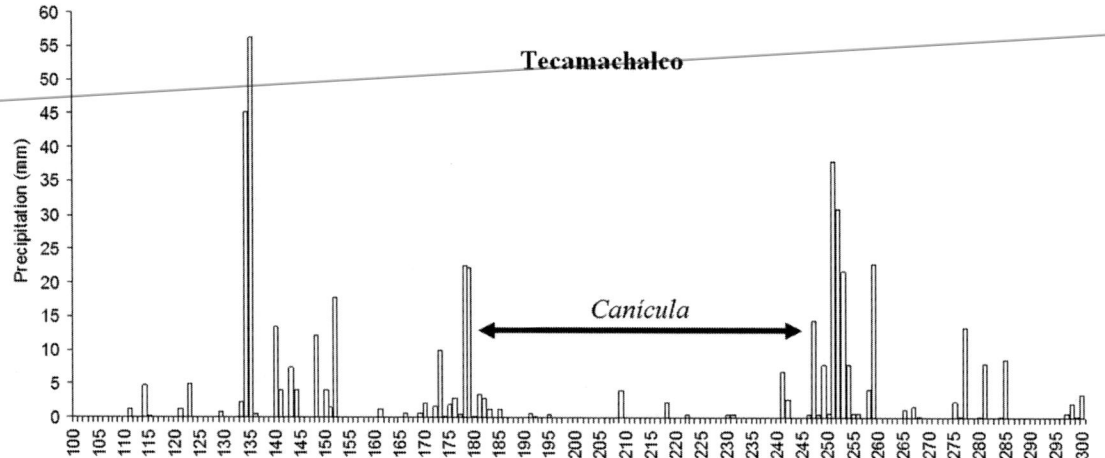

FIGURE 46. DAILY PRECIPITATION RECORDS FROM DAY 100 (APRIL 10 OF 2009) TO DAY 300 (OCTOBER 27 OF 2009) FOR THE TECAMACHALCO WEATHER STATION.

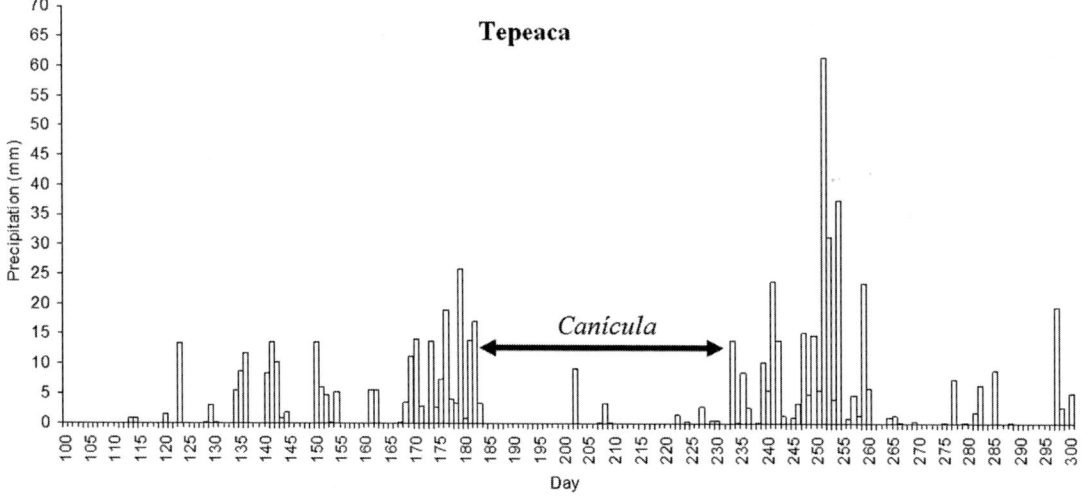

FIGURE 47. DAILY PRECIPITATION RECORDS FROM DAY 100 (APRIL 10 OF 2009) TO DAY 300 (OCTOBER 27 OF 2009) FOR THE TEPEACA WEATHER STATION.

Agricultural Production for the Year 2009

Precipitation (mm)					
	2005	2006	2007	2008	2009
February	0.2	0	55.4	0.2	2
March	4.4	0.8	2.4	0.2	8.6
April	43.2	8	67.6	23.2	3
May	66.6	13.6	111.2	39.4	97
June	25	53	59.4	198.8	143.6
July	91.8	117.2	148.4	118.6	33
August	93.2	129.2	129.2	88.4	86
September	59.4	100.8	161.4	194.4	217.4
October	54.4	59.4	61.8	3.8	53.4
November	7.6	44	18	0.2	1
December	3.8	1	2.8	0.2	7.4
Total	449.6	527	838	667.6	654.2

FIGURE 48. PRECIPITATION RECORDS FROM 2005 TO 2009 FOR THE TEPEACA WEATHER STATION. DATA OBTAINED FROM INIFAP

Precipitation (mm)				
	2006	2007	2008	2009
January	0.2	24.8	0.4	0.2
February	0	31.4	0	6.2
March	15.2	4	2.2	2
April	34	60.2	7.8	6
May	212.8	122.2	38.8	157.2
June	101.8	117.6	72.2	87.2
July	63.8	88.4	86.6	10.4
August	71.8	67.6	50.2	12.8
September	66	67	46.2	153
October	36	95.8	29	58.6
November	47	5	0.2	0
December	0.6	1	0	3.8
Total	649.2	685	333.6	497.4

FIGURE 49. PRECIPITATION RECORDS FROM 2006 TO 2009 FOR THE TECAMACHALCO WEATHER STATION. DATA OBTAINED FROM INIFAP.

The canícula drought was most severe across the southern portion of the Tepeaca valley. In the area from Cuauhtinchan to Tecamachalco, the agricultural cycle normally starts from May to early June. INIFAP's website data (www.inifap.gob.mx) indicates that in 2009, precipitation was good in these two months ranging from 87 to 157mm (Figure 49). Yet by July and August, precipitation had fallen to only 10 to 13mm. Here, the lack of water did not affect initial growth or germination; it affected plant growth, inflorescence and ear development. When visiting the fields in late august, farmers told me they still had hopes that rain would fall in time to provide water for ear development. 'If rain does not come within 15 days, the cultivars will be lost', one of them told me. Unfortunately for them, the rains did not arrive.

INIFAP's (www.inifap.gob.mx) records show that in September, it rained six times (153mm) more than in July and August combined (23.2mm). This was almost four times the amount of rain for the same month in 2008 (46.2mm), and twice as much as in 2006 (67mm) and 2007 (66mm). In the southern portion of Tepeaca valley and the eastern part of the Llanos de San Juan, not only was total precipitation for 2009 relatively low (497mm), but it was unequally distributed. In the Tecamachalco region 310mm (62%) was concentrated in May and June while July and August accounted for only 23.2mm (5%). At Atenco, July and August only had 72mm (17%) of the annual total of 433mm, while September had 230mm (53%). Acatzingo had a similar pattern; the yearly total was 669mm, of which only 91mm (14%) fell in July and August, compared to 242mm (36%) in September.

Maize is very sensitive to stress situations, especially with regards to moisture deficiency during the female and male inflorescence periods (Maiti and Wesche-Ebeling 1998). This was exactly the problem that farmers faced in the 2009 season. Figure 50 shows the time line in which maize cultivation took place in two fields near Acatzingo in relation to precipitation records. The graph clearly shows that by the time of male and female inflorescence development, there was a clear water deficiency.

Comparisons from eight different weather stations throughout the study area show that the 2009 lull in rainfall did not strike equally. The southern area was most severely affected while the least affected one was the western part of the Puebla-Tlaxcala valley. Paradoxically, the city of Puebla weather station recorded pretty much what a typical rainy year might look like. Its mean precipitation was of 816mm and monthly rainfall had a bell-shaped curve distribution (Figure 51).

Only Household 23 had fields in two different environmental areas. Two were in the Tenango area in the Tepeaca Valley, and one more near Santa Ana in the Llanos de San Juan Region. This third field is located 7km from the household, but since the Tezontepec-Cuesta Blanca mountain range lies between them, members of Household 23 have to some travel 20km around these mountains to reach this field. Cultivating the Santa Ana plot was feasible only because an arrangement was made with another household from Santa Ana to work it 'a medias' (by half), that is, one owner provides the land and the landless partner supplies the labor. At harvest the crop is shared equally between both. Also, Household 23 possessed a truck that could be used to visit the field. Although one might expect that their production ought to be better because they planted more fields than other households, and in different environmental zones, their production was still low (517kg/ha) (Figure 52). Note that the production of Field 2 was split up between two household and thus only represents 176kg/ha for each

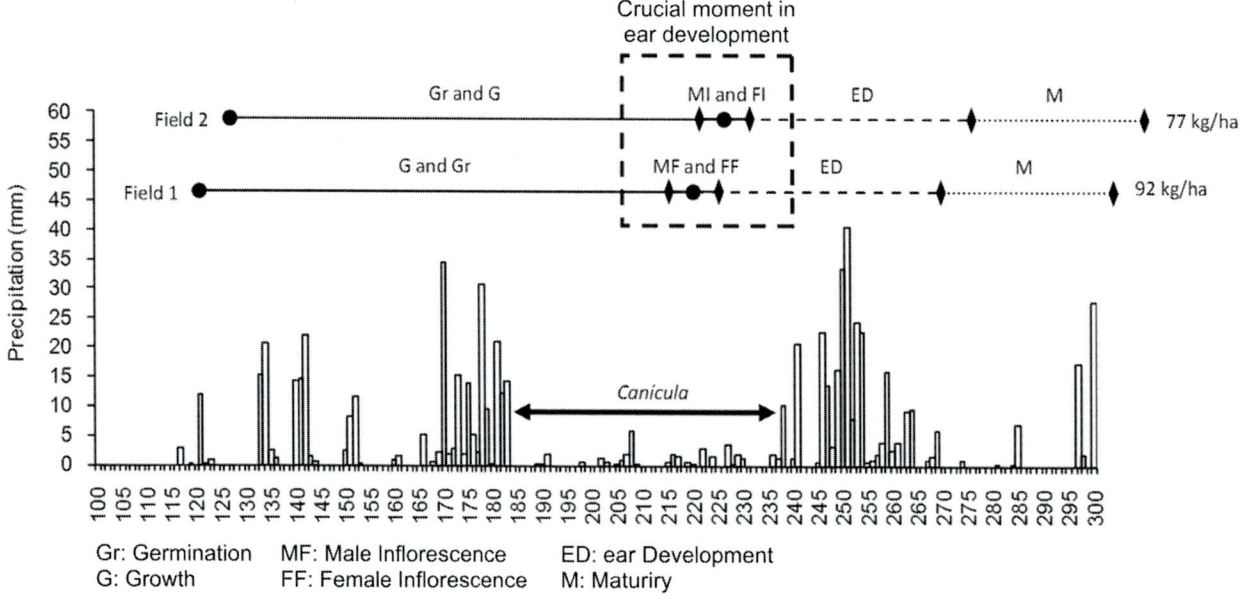

FIGURE 50. 2009 AGRICULTURAL CYCLE FOR MAIZE IN TWO FIELDS FROM HOUSEHOLD 17 IN RELATION TO PRECIPITATION IN THE ACATZINGO AREA.

household. Therefore, cultivating this field resulted in little gain, but it nonetheless was welcomed because relatively few resources were inputted (gasoline for the truck and some labor).

It would be expected that under normal climatic regimes, field dispersion may act as a buffering strategy against total crop failure. If the drought was localized, having two or more fields in different areas would have lessened its effect. In other words, if one field failed the other would still be available.

The data suggests that field dispersion did not help to ameliorate the effects of the dry canícula. To establish if this was indeed the case, a correlation analysis was conducted in order to establish the possible relationship between field distancing and maize production. For each household, I obtained the distance between a set of two fields and averaged their maize yields (Figure 53). With these two set of values, I then ran the correlation analysis.

The result shows a weak correlation ($R^2 = 0.24$) between field dispersion and average maize yields. Only in a small number of cases did greater field separation result in better average yields for the household. In most cases it did not matter if fields were distanced closer or further away from one another because yield values varied considerably. At least for the year 2009, the strategy of field dispersion was insufficient for coping with the extended water deficit brought about the canícula. Thus, under the prevalent rainfed system, farmers had no way to fight this problem.

Field dispersion is widely used in the study region as a strategy for averaging maize production. One thing to keep in mind, however, is that farmers use field dispersion mainly to average yields and, in some way, to increase overall production. Distancing fields can buffer risk of total crop failure, but generally farmers cannot surpass a certain level of productivity. This is especially true for households that have fields in poor environmental conditions, such as Households 17 (Acatzingo) and 24 (Cuapiaxtla) located in the southeast portion of the Tepeaca valley. In Household 17, it's Fields 1 and 2 were set 1784m apart. On the contrary, Fields 1 and 2 of Household 24 were distanced one another much further at 4853m apart. In spite of this variability in field separation, average maize yields from these two households were almost identical yielding only 67 and 85kg/ha respectively.

In contrast, the fields of Household 14 (Soltepec), located in the Llanos de San Juan, were also dispersed at variable distances of 2466m (fields 1 to 2), 5693m (fields 1 to 3), and 4431m (fields 2 to 3). Average maize yields for each set of fields were 645kg/ha, 892kg/ha, and 560kg/ha respectively, also showing no consistent trend.

From the results of this exercise it appears that field dispersion is not a useful strategy to buffer a severe regional climatic disruption. Under normal conditions, field dispersion might mitigate the negative effects of environmental unpredictability and help increase the amount of maize available for the household by averaging overall yields. Yet, regional droughts such as the 2009 canícula can affect large areas far beyond the distance in which smallholders locate their fields. This inevitably leads to a considerable lowering in average yields no matter the distance at which fields are dispersed.

Other negative phenomenon

The previous analysis of field dispersion and total production suggests that field dispersion is directed

Agricultural Production for the Year 2009

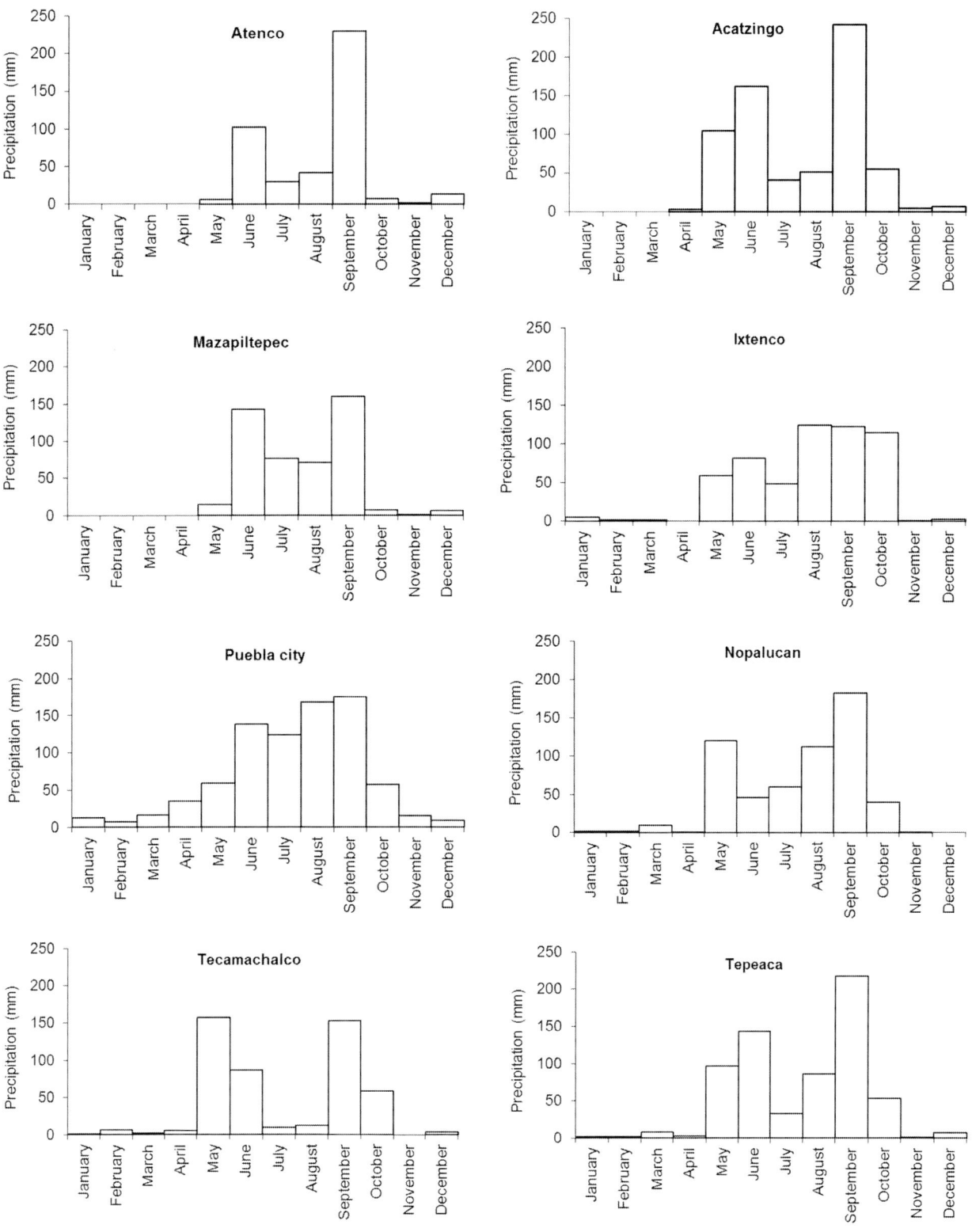

Figure 51. Monthly precipitation records from eight weather stations from the Tepeaca, Llanos de San Juan and Puebla-Tlaxcala valleys during 2009.

Household 23	Field 1	Field 2 (*a medias*)	Field 3
Maize yields (kg/ha)	103	352 / 2 households	154
Cultivated surface (ha)	0.50	0.84	1.08
Total production (kg)	52	299	166
Total maize production for the household: 517 kg			

Figure 52. Maize production for Household 23 from Tenango for the year 2009.

Household	Fields	Distance between fields (m)	Average yields (both fields kg/ha)
1	1 to 2	2127	421
2	1 to 2	5269	403
2	1 to 3	2313	601
2	2 to 3	2997	566
3	1 to 2	1603	97
4	1 to 2	2776	96
5	1 to 2	1791	73
7	1 to 2	1838	75
8	1 to 2	2118	54
8	1 a 3	1626	116
8	2 a 3	3542	97
9	1 to 2	1634	93
10	1 to 2	2852	426
11	1 to 2	1988	533
14	1 to 2	2466	645
14	1 to 3	5693	892
14	2 to 3	4431	560
15	1 to 2	2580	852
16	1 to 2	1995	494
17	1 to 2	1784	85
18	1 to 2	3136	130
19	1 to 2	4022	306
19	1 to 3	3046	235
19	2 to 3	1914	260
20	1 to 2	2676	88
21	1 to 2	5776	507
21	1 to 3	4904	471
21	2 to 3	1240	341
22	1 to 2	3843	604
23	1 to 2	6782	229
23	1 to 3	3395	129
23	2 to 3	6269	254
24	1 to 2	4853	67

Figure 53. Distances between fields and average yields per household.

primarily towards buffering minor climatic events and other negative phenomena. These would include pests, typical precipitation variability, theft, and differential soil quality.

In general, pests and seed predation can considerably reduce total productivity. Interestingly, farmers expect that some proportion of seeds will not germinate or be lost to predators. This is part of the reason why they tend to sow three to four maize kernels and two or three bean seeds together. Some of the seeds will end up as food for rodents, birds and insects, while others may simply not have enough vigor to sprout. Certain insects like miners affect plants during their early developmental stages. Usually, these infestations have low-level effects and are simply tolerated because there is little that can be done to fight it. When infestations reach a higher level, some farmers will buy pesticides at local suppliers, but this is normally only done in fields where cash crops are raised.

Even with infestations, it is expected that three out of four kernels will survive to maturity (75% to 80%), although in some cases it may be less. When the problem reaches unacceptable levels, some portions of the plot may be re-sowed.

In 2009, pest infestation, predation and theft were not major problems for agriculturalists. Yet, during the second half of the season, after the drought period, heavy rainfall stimulated enormous weed infestation. This was not good for the small number of plants that were able to survive and develop ears or pods. Plant competition for nutrients became stronger. In any other year, farmers would have made an effort to remove weeds, but since they knew yields would be low, they sought governmental aid by reporting total crop failure and made no effort to clean the plots. This resulted in weeds taking over crop stands.

Several people were seen pasturing goats, sheep and cows in fields late in the cycle, after stalks were dried out. Generally, pastoralists enter fields after the owner has cleared away the upper portions of the maize stalk and has left only the lower part of the stalk with the roots called '*gallos*' (roosters). But sometimes, they will enter the fields earlier if they consider that the owner has no intention or interest in cutting down the corn stalks. Although this could be considered as invasion or theft, farmers do not make a big deal of it and in general consider it a common practice. As long as only a small fraction of the maize stalks are consumed it is considered as a favor, performing a good deed and lending a hand to other people.

The heavy downpours that struck the region after the dry canícula also promoted the growth of corn smut fungus (*Ustilago maydis*) locally known as *huitalcoche* or *cuitlacoche*. This fungus is edible and is consider one of the most traditional foods in the region. Although it's not necessarily desired, it does have the advantage that it is sold at local or city restaurants. It occurs when ears of maize are infected by the basidiomycete *Ustilago maydis* fungus which produces young, fleshy and edible galls (Pataky and Chandler 2003). Although I did not measure its incidence, farmers of some communities assured me that there was much more *cuitlacoche* in 2009 than in previous years.

Although I did not observe any field with pest infestation, the September rains brought about a proliferation of chapulín grasshoppers in some areas of the piedmont around the La Malinche volcano. These insects can be a serious plague for crops because they attack the leaf systems, notably reducing photosynthesis functions. Yet, chapulín are an important protein food source and are consumed regularly in some communities. There are certain households that profit from this insect plague and will collect and prepare grasshoppers for their own consumption or to sell at local *tianguis* or marketplaces.

Work inputs to field

Asking about the amount of work put into the fields this year seemed to be an odd question for most farmers. They consider that if a person cultivates a field, it is implicit that he/she is willing to carry out all the tasks that are needed throughout the year. It is only under exceptional circumstances that someone will abandon the field without completing the necessary work. Valid reasons include things such as sickness, accidents, legal problems, migration, death or severe climatic events such as happened in 2009. But even in these cases, there is generally sufficient labor available for a household to balance the absence of one person. In fact, it is very common that work is divided between household members. For example, the barbecho work is a male's job, while weeding may be done by women and children. If for any reason someone is not able to do certain tasks help may be asked from kinsmen (a father, brother, cousin or an older son) or a friend with the arrangement that it will be compensated later on.

Even though many farmers abandoned their fields as a result of the canícula, a few tried to contend with these problems. For example, Household 21 had fields located close to the Acajete Plateau where tremendous problems with rainfall occurred. The effect of the canícula was slight and for most of the year rains were constant and strong. This forced the owners to work harder than previous years in order to avoid the compacting of the soil in flooded lands. Apparently, most fields in this area were affected by rain and abrupt downpours, resulting in many farmers abandoning their fields. Fields in this area lost their productivity due to tremendous weed development and sometimes the decomposition of plants. I cannot explain why many decided to abandon their fields, but it is probable that many of them pursued other economic activities that provided them with more resources than the fields. However, the head of Household 2 was too old and did not have any other activity to engage in. The fields were in fact primordial for his food resources.

Types of rain

Soil and water retention directly influence sowing densities. Although soil properties and texture influence moisture content, fields that depend on rainfall usually have some water retention issues because potential evaporation normally exceeds precipitation. This is especially true for areas in the southern portion of the Tepeaca Valley such as Tecali and Hueyotlipan. Therefore, the soil's capacity for retaining moisture is an important variable for crop production. No matter how much it rains, if the soil has poor water retention, there will be little moisture available for the development of the plants.

When questioning farmers about the type of precipitation that they preferred, invariably they all responded that light and steady rain was the optimum for water retention in the soil. These rains are associated with coastal frontal systems and hurricane storms striking the Gulf or Pacific coasts. Seasonal summer rain, for most part is intense and rapid, usually lasting from a couple of minutes to half an hour. Precipitation is fast and copious, and although it is much welcomed, it may not be adequately absorbed by the soil and may runoff or cause soil erosion problems. This will obviously depend on the degree of absorbency of the soil. Sandy soils have the best absorbency while clayish soils had the most problems with water repellency.

On the other hand, frontal storms producing steady and light rain and can last for several days depending on hurricane events in both the Pacific and Atlantic oceans (Magaña *et al.* 2004: 47-55). This rain is intermittent allowing ample time for the soil to absorb water. In areas with steep gradients, erosion may be a problem if the rains continue for too many days. Similarly, valley bottoms can be affected by floods. Moisture content in the deeper alluvial areas is thoroughly replenished by these rains and can remain in the soils for extended periods lessening fluctuations in water supply to the crops. However, when steady rains occur at a continuous rate, soils of lesser quality also can be greatly benefited.

Second part: regional agricultural productivity in the Tepeaca region

Regional maize productivity can be expected to vary within the Tepeaca region. Soil quality, precipitation levels, humidity retention in the subsoil, and evaporation-transpiration rates are the four main variables that condition agricultural productivity (Instituto Nacional de Investigaciones Forestales 1997). These variables can be combined to establish an ordinal scale of five classes of potential zones for agriculture (Figures 54 and 55). Class 1 zones have the best conditions for agriculture and are found in Acajete, Amozoc la Joya, Tlaxco, Tepatlaxco and Zitlaltepec. Class 2 zones are found in the central and northwestern corner in and around the Piedmont areas of the La Malinche Volcano embracing areas like Macuilá, Nopalucan, Tenextepec and Tlayoatla. The southwest and northeast areas have predominantly Class

Class	Soil type and quality		Humidity retention in the soil	Evapo-transpiration (mm)	Precipitation (mm)
1	Be+Je+Hc/1	I+E+Be/2	9 to 10 months	1400-1900	800-900
2	Be+Bk+Vc/2	I+E+Be/2	8 to 12 months	1400-1900	600-1000
	Be+Hh/1	Je+Bv/1			
	Be+Je+Hc/1	Je+Lo/1			
	Hh+Be+I/2/D	Th+Rd/1/P			
3	Be+Bk+Vc/2	I+E+Be/2	6 to 12 months	1400-2000	500-1000
	Be+Hh+To/2	I+Rc+E/2			
	Be+I+E/2/L	Je+Bv/1			
	Hh/2	Rc+Bk/2/L			
	Hh+Be/2	Re+I+Hh/2/L			
	Hh+Be+I/2/D	Th+Rd/1/P			
	I/2	Th+To/2/L			
	I+E/2				
4	Be+Hh+To/2	Rc+Bk/2/L	5 to 9 months	1500-2000	500-800
	Be+I+E/2/L	Re+I+Hh/2/L			
	Be+To+Th/2	Th+To/2/L			
	Hh/2	Vc/3			
	I+C/2	Xh/2/PC			
	I+E+Be/2	Xh+Hc/2/PC			
	I+Rc+E/2				
5	Be+To+Th/2	Xh/2/PC	5 to 8 months	1700-2000	500-700
	Re+I+Hh/2/L	Xh+I/2/PC			
	Xh/2				

FIGURE 54. CLASSIFICATION OF THE DIFFERENT AGRICULTURAL POTENTIAL ZONES WITHIN THE STUDIED REGION. FOR THE SOIL NOMENCLATURES SEE FIGURE 9.

3 and 4 zones and include Cuauhtinchan, Mazapiltepec, Soltepec, Tecali, Tenango and Tepeaca. Class 5 zones are the least useful for agricultural practices and are located mainly in the southeastern portion in Mixtla, Cuapiaxtla, Hueyotlipan, Caltenco and Tecamachalco, and in a small area near Atenco in the northeastern corner. The areas above 2700m in altitude are used sporadically for maize cultivation. Above such altitude environmental conditions become too harsh and cold for maize cultivation. Some long developing maize varieties do tolerate such climatic conditions and are sporadically employed in the high piedmont of the La Malinche volcano (Lara *et al.* 2002). However, given that these are small patches within the studied region, these high altitude maize cultivars were not consider in this study.

Today, average maize production under rainfall conditions is moderately good within the studied area. Under subsistence agriculture one metric ton of maize is considered the minimum amount required annually for an average household of 5 to 7 individuals with an 80% dependence on maize (Sanders *et al.* 1979: 372-373, Table 1, Steggerda 1941). Below this amount, maize productivity is insufficient and households must acquire food through other means.

Achieving this amount will depend on the size of cultivated area and the amount of maize produced by unit of space. It would be expected that in an average year Class 1 zones would produce higher maize yields and households would need to work less land to obtain the necessary supplies. As one moves down the classification, maize productivity should decline and households would need to cultivate more land or acquire food through other means.

The distribution of potential class zones correlates well with historical maize yields within the study region (Figure 56). According to SAGARPA´s website (http://www.oeidrus-df.gob.mx/aagricola_dfe/ientidad/index.jsp), maize yields by municipio averaged 1434kg/ha, with a standard deviation of 1080kg/ha (Figure 57). This average is similar to the worldwide average maize production (1232kg/ha) presented in Figure 5. However, yields differ according to each environmental region (Figure 58). Fields in the northern Llanos de San Juan located in Class 1, 2 and 3 zones report averages of 2222kg/ha over the past eight years. The uppermost averages have been recorded in San Salvador el Seco in 2008 (3500kg/ha), Aljojuca in 2008 (3200kg/ha), and Mazapiltepec in 2001 (3200kg/ha). The western portion that includes the Puebla-Tlaxcala valley has predominantly Class 1 and 2 zones and has reached the

AGRICULTURAL PRODUCTION FOR THE YEAR 2009

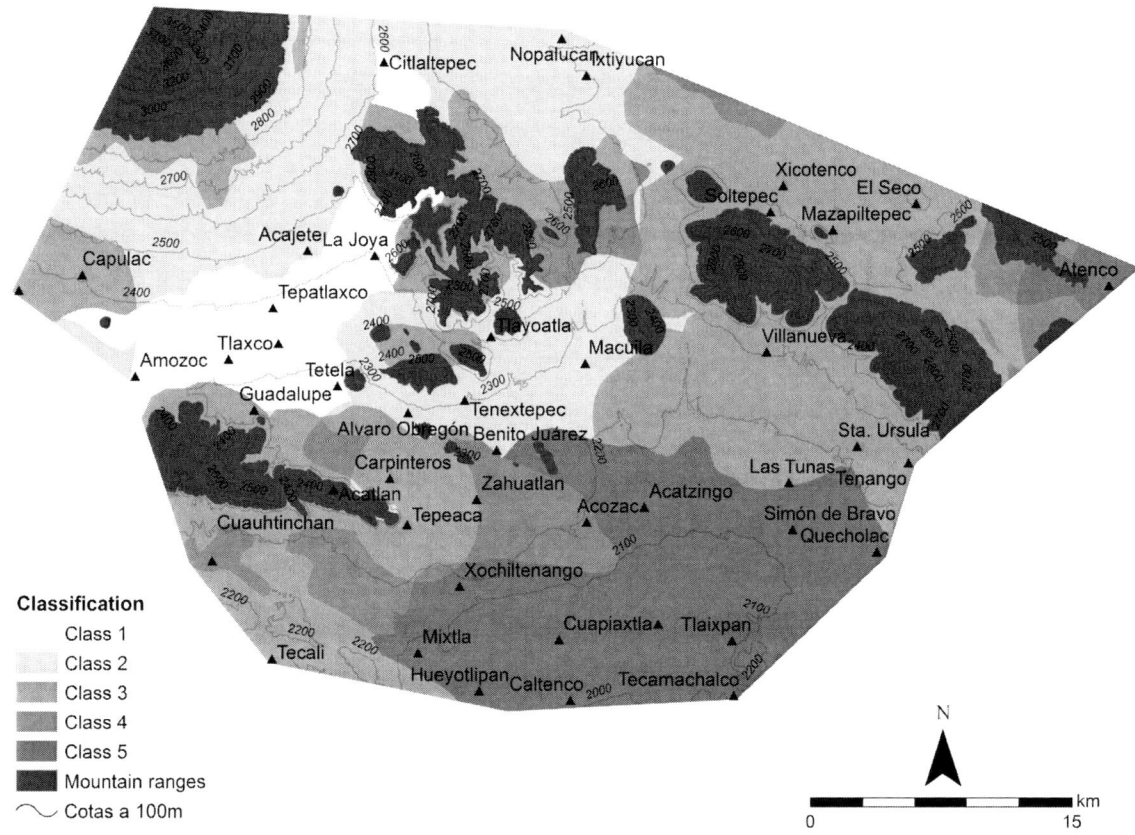

FIGURE 55. MAP SHOWING THE CLASSIFICATION OF DIFFERENT POTENTIAL ZONES FOR AGRICULTURE.

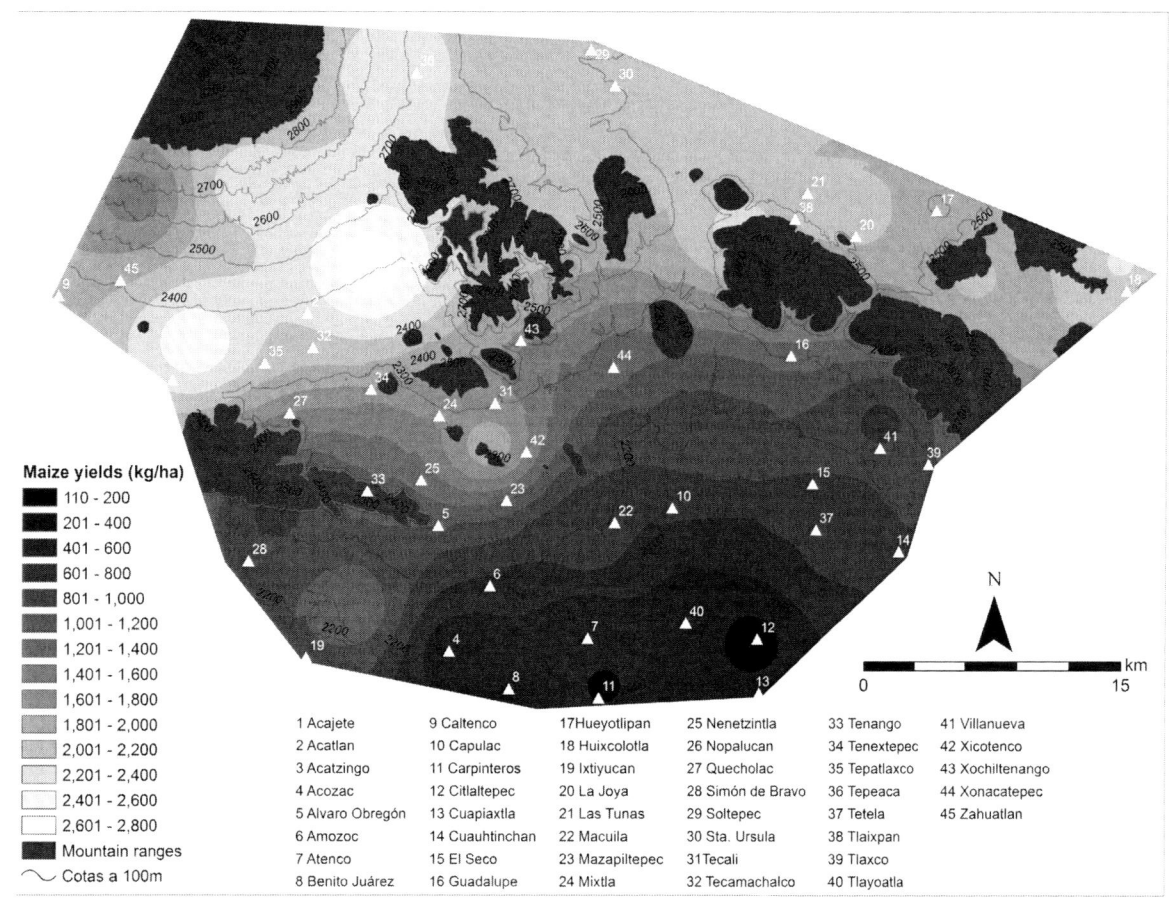

FIGURE 56. AVERAGE MAIZE YIELDS BY MUNICIPIO FOR 2001-2008 WITIN THE STUDIED REGION.

Maize yields per year (kg/ha)									
Municipio	**2008**	**2007**	**2006**	**2005**	**2004**	**2003**	**2002**	**2001**	**Average**
Acajete	3000	3300	3000	3000	3000	2600	1688	2500	2761
Acatzingo	2000	1200	1500	668	800	509	423	1500	1075
Aljojuca	3200	2940	3000	2500	905	2000	923	3000	2308
Amozoc	3000	3300	3000	3000	3000	2000	1909	3100	2789
Cuapiaxtla de Madero	444	650	800	131	25	97	15	25	273
Cuautinchan	800	800	800	800	800	500	1420	800	840
General Felipe Angeles	1490	1280	1200	340	1100	125	175	1090	850
Mazapiltepec de Juarez	3500	2306	3000	958	2000	2318	361	3200	2205
Mixtla	500	650	500	163	24	90	0	74	250
Nopalucan	3200	2500	3000	812	2000	2154	1250	2732	2206
Puebla	2125	3000	2000	813	813	1055	1399	2400	1700
Quecholac	1471	1194	1100	200	762	95	120	660	700
San Juan Atenco	3000	2123	3000	2000	1632	2419	461	2688	2166
Reyes de Juarez, Los	1000	1350	1700	164	219	318	67	1500	790
San Salvador El Seco	3500	2498	2500	998	2000	1833	1971	2400	2213
San Salvador Huixcolotla	800	800	747	164	56	147	0	51	346
Santo Tomas Hueyotlipan	800	800	750	143	53	145	45	65	350
Soltepec	3500	3000	3000	533	2000	2455	1250	2274	2251
Tecali de Herrera	1500	1600	1500	1500	1500	275	1200	1100	1272
Tecamachalco	400	85	250	146	11	0	15	115	128
Tepatlaxco de Hidalgo	3000	3200	3000	3000	3000	1440	1663	2400	2588
Tepeaca	2000	2500	2000	2000	2000	1292	1499	2400	1961
Tochtepec	464	450	159	320	0	0	0	100	187
Zitlaltepec	2950	2500	2000	1100	2410	2410	1940	2400	2214
Average	**1985**	**1834**	**1813**	**1061**	**1255**	**1095**	**825**	**1607**	**1434**

Figure 57. Average maize yields by municipio for 2001-2008 within the studied region. Source: SAGARPA.

Region	Municipios	Average maize yields (kg/ha)	Range
Llanos de San Juan	Atenco, Aljojuca, El Seco, Huamantla, Mazapiltepec, Nopalucan, Rafael Lara Grajales, Soltepec, Zitlaltepec	2222	361-3500
Puebla-Tlaxcala valley	Acajete, Amozoc, Puebla, Tepatlaxco	2459	813-3300
Tepeaca Valley	Acatzingo, Cuapiaxtla, Cuauhtinchan, Felipe Ángeles, Mixtla, Quecholac, Reyes de Juárez, Huixcolotla, Hueyotlipan Tecali, Tecamachalco, Tepeaca, Tochtepec	694	0-2500

Figure 58. Average maize yields according to environmental region within the study area. Data from SAGARPA for the years 2001-2008.

highest average of 2459kg/ha within the study region. In particular, the Amozoc, Acajete and Tepatlaxco areas with Class 1 lands, are by far the most productive and stable of the region usually averaging more than 3000kg/ha.

In contrast, the municipios in the Tepeaca valley include principally Class 3, 4 and 5 zones. These areas have markedly low maize average yields of 694kg/ha. In particular, the southern portion has a dramatically low productivity with values usually below 300kg/ha and commonly reported as total crop failure (a value of zero). Excluding the zero values, the lowest yields correspond to Tecamachalco in 2003 (0kg/ha), Mixtla in 2004 (20kg/ha), Hueyotlipan in 2005 (140kg/ha), and Tochtepec in 2006 (160kg/ha).

The 2009 maize production in the study region

In general, maize production for the year 2009 was poor throughout the study region. The major problem was the prolonged drought produced by the dry canícula which altered precipitation in many areas. This unfortunate event became an excellent opportunity for analyzing the impact of such a climatic event at a regional level. This required that the sample be broadened beyond the initial 51 fields from the 24 households surveyed. In doing so, more fields were included in the survey from more areas with different environmental settings.

This dimension of research was not considered in the initial study planning and was centered on answering the question of how heterogeneous the effects of the canícula were on crop production throughout in the entire region. If the effects were not homogeneous, which areas were affected more or less and to what degree? This piece of information looked promising for examining variability in production as a crucial element for reconstructing ancient Postclassic maize production and understanding household risk. It would also provide information on how noble elites might have managed their estates and why staple extraction from commoner households was organized the way it was within their domains.

This second survey also helped to correct imminent problems that arose with data collection during the household survey. One problem was the restricted area over which the 2009 production estimates were obtained. Because the sample was small in relation to the large area being studied, to understand these effects I needed a broader sample of how the canícula affected fields that differed from one another in soil type, weather, slope, moisture retention, and other key regional characteristics. This was needed in order to do comparisons and potential correlations between soil quality, altitude, weather and agricultural production. The idea was to observe if production was higher in those areas considered better for agriculture, or if precipitation and moisture retention properties of soil were the prominent variables when rainfall shortages occur during the dry canícula.

Methodology

For the larger survey I examined an additional 449 plots. The only reason for this is that I arbitrarily wanted to reach the 500 fields measured. Although this was a relatively high number, the methodology developed for estimating total production made it simpler and relatively easy to calculate yields. The main constraints were time and the distances needed to travel. Nevertheless, in the end this proved to be a very insightful and useful decision.

The 449 fields were selected through random point generation and using the historical 16th data of the Tepeaca *altepetl* as the boundary of the survey (Figure 59). Here, there was no stratification of any sort, only a simple random sample. Areas over 2700m where excluded from examination because, as it was mentioned before, maize is only sporadically cultivated above this altitude. These areas comprise mainly the higher areas of the La Malinche volcano and the Tezontepec and Cuesta Blanca mountain ranges.

The 449 fields were visited from September to December of 2009 and were located on a map using a GPS. Production calculations were made using the Ear Volume Method and plant sowing densities and followed the same methodology described for the household survey. For calculating production variation in the study region, I combined all the information collected from the 51 fields from the household survey with the 449 fields from the expanded sample. The values for each agricultural plot were transformed into the volume of shelled maize per hectare and its equivalent measurement in kilograms (Appendix A). The data was processed and plotted on a map using spatial interpolation analyses. Spatial interpolation is a useful type of spatial prediction of '… the exact values of attributes at unsampled locations from measurements made at control points within the same area' (O'Sullivan and Unwin 2003: 220). The interpolation method favored here was the Inverse Distance Weighted approach in which cell values are estimated by averaging the values of sample data points in the neighborhood of each processing cell. Nearer locations are given more prominence or weight in the calculation of a local mean. The interpolation used a default variable search radius of 12 points or neighbors and a significance power value of 2 (Figure 60).

The interpolation produced a smooth regional maize production map clearly showing the pattern of effects produced by the 2009 canícula event. The results from the survey indicate that the average maize production for the region in 2009 year was 371kg/ha. The minimum yield was 30kg/ha and the maximum was 1108kg/ha. It seems that both altitude and latitude have a strong weight on regional maize production. In general, higher elevations with a cooler and more humid climate are found in the northern portions of the study region. The area around the La Malinche volcano is especially wetter due to the rainfall effects caused by the volcano and the deeper soils of the area (see Figure 10). The south is both lower in elevation and has drier conditions. The SAGARPA data for the years 2001-2008 indicate that the Llanos de San Juan located in the north tend to have high production levels. This pattern was not altered by the 2009 dry canícula and higher yields were correlated with higher elevations in the northern latitudes (Figures 61-64). Altitude may be the main cause that influences the spread of the dry canícula drought event. As one moves north and west into higher altitudes, the canícula effect weakened and productivity increased. The effect of the canícula was least on field locations between 2250 and 2600m while the effects of the drought were greatest on fields below 2200m.

Maize yields in the study region according to land class

A comparative analysis was made regarding maize yields according to land classes within the three major regions inserted in the study region. The data recorded from the 2009 fields surveyed was organized according to land

Rainfed Altepetl: Modeling Institutional and Subsistence Agriculture

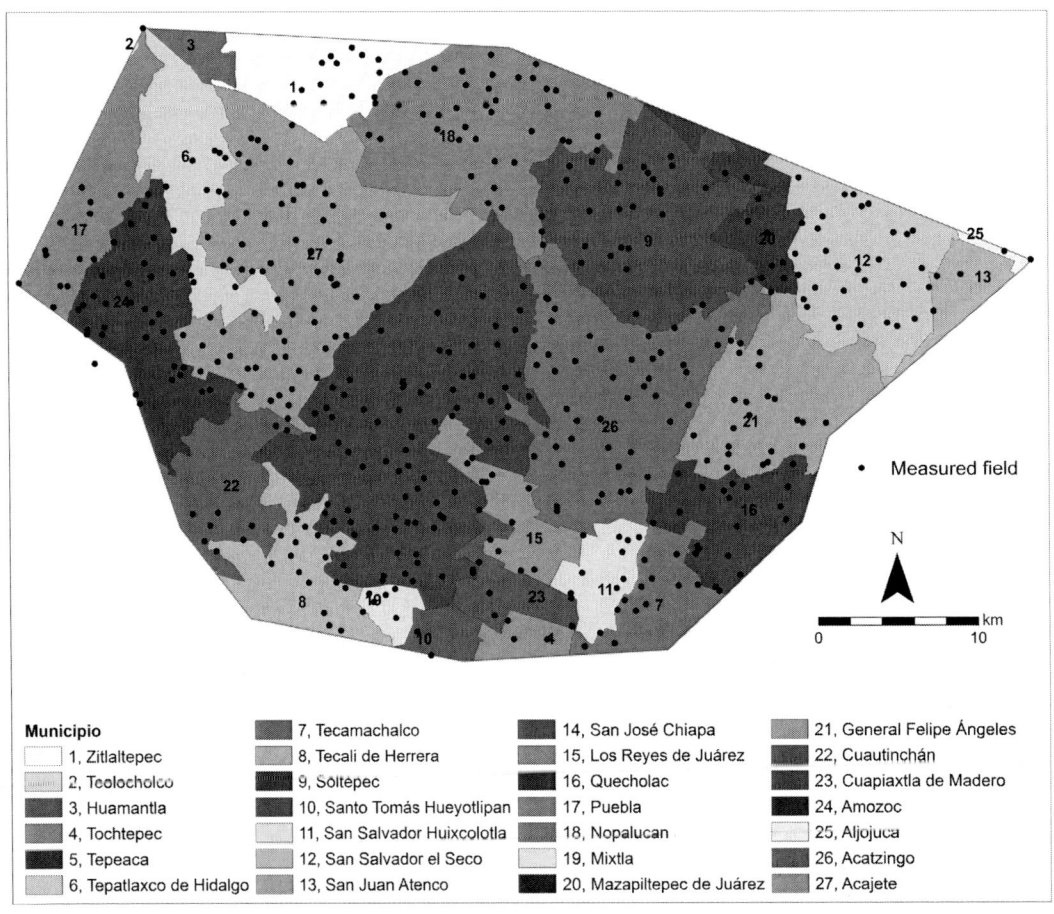

Figure 59. Distribution of the 449 fields measured plus the 51 fields of the household survey within the different municipio boundaries.

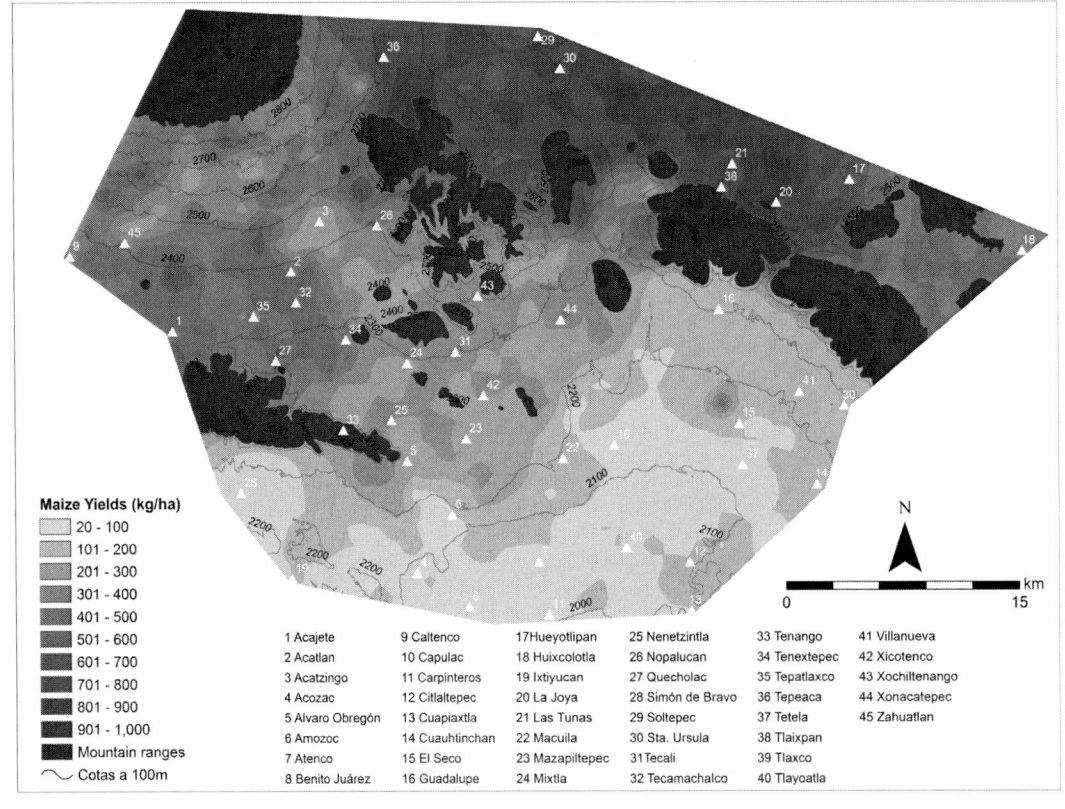

Figure 60. Map showing maize production within the Tepeaca Region for the year 2009 based on the total of 500 fields and intervals of 100 kilograms per hectare.

Model Summary and Parameter Estimates							
Dependent Variable: Maize Production							
Equation	Model Summary					Parameter Estimates	
	R Square	F	df1	df2	Sig.	Constant	b1
Linear	.367	289.036	1	498	.000	-1893.071	.972
The independent variable is Altitude							

FIGURE 61. RESULTS OF THE REGRESSION ANALYZES RUN IN SPSS SOFTWARE FOR MAIZE YIELDS IN KILOGRAMS PER HECTARE AND THE ALTITUDINAL LOCALIZATION OF THE FIELDS.

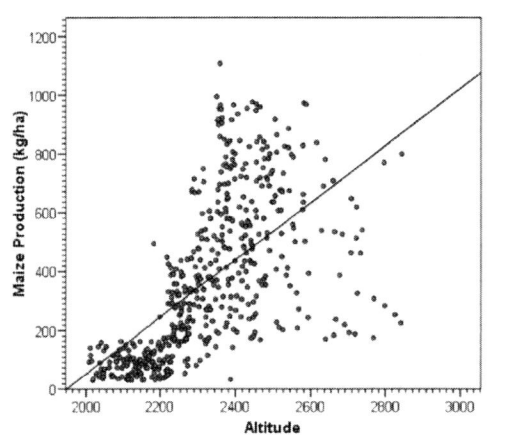

FIGURE 62. RESULTS OF THE REGRESSION ANALYZES RUN IN SPSS SOFTWARE FOR MAIZE YIELDS IN KILOGRAMS PER HECTARE AND LATITUDE OF THE FIELDS.

Model Summary and Parameter Estimates							
Dependent Variable: Maize Production							
Equation	Model Summary					Parameter Estimates	
	R Square	F	df1	df2	Sig.	Constant	b1
Linear	.579	683.588	1	498	.000	-45460.7	2405.798
The independent variable is Latitude							

FIGURE 63. REGRESSION ANALYZES FOR MAIZE YIELDS AND ALTITUDE.

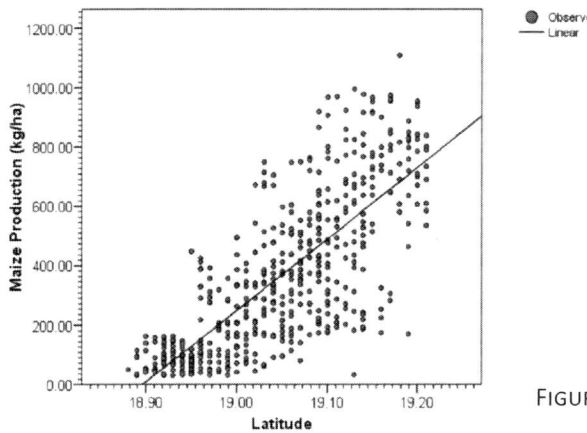

FIGURE 64. REGRESSION ANALYZES FOR MAIZE YIELDS AND LATITUDE.

Land Class	Tepeaca valley		Llanos de San Juan		Puebla-Tlaxcala valley	
	Range	Average	Range	Average	Range	Average
1	-	-	798-886	838	175-784	489
2	194-441	305	171-960	732	175-800	407
3	34-679	209	350-1108	681	182-799	476
4	41-448	206	426-752	562	-	-
5	30-437	94	399-558	482	-	-

FIGURE 65. RANGE AND AVERAGE MAIZE YIELDS (KG/HA) DURING THE 2009 CANÍCULA ACCORDING TO LAND CLASS WITHIN THE STUDY REGION.

classes present within the major region inside the Tepeaca region. The information on maize yields ranges and averages is presented in Figure 65.

The Tepeaca valley

Within the Tepeaca valley there are important differences in maize yields according to land classes (Figure 65). Fields located in the best Class 2 lands in Macuilá, Tetela, and Tlayoatla yielded higher than the rest of the valley and had a range of 194-441kg/ha with a mean of 305kg/ha. Yields in Class 3 lands, located on the west and east portions of the valley were also strongly affected and yielded between 34 and 679kg/ha with an average of 209kg/ha. The lowest yields within the valley occurred in the south and southeast portions were marginal conditions prevail. The canícula intensified the already marginal conditions of Class 5 lands in Caltenco, Cuapiaxtla, Huixcolotla, Mixtla and Tecamachalco, where maize yields ranged from 30 to 176kg/ha and the average was of only 88kg/ha.

The Puebla-Tlaxcala valley

The Puebla-Tlaxcala valley has the best conditions for agricultural production with Class 1 to 3 lands. However, in 2009 maize yields varied considerably (Figure 65). Fields in Class 1 lands had an average of 489kg/ha with a range of 175 to 784kg/ha. Class 2 and 3 lands had a polarized range and a low mean. Class 2 lands produced between 175 and 800kg/ha with a mean of 407kg/ha. Class 3 lands produced an average 476kg/ha with a range of 182 to 799kg/ha.

The Llanos de San Juan

Important differences in productivity were observed between locations within the Llanos de San Juan (Figure 65). Class 1 lands, located in a small area in Zitlaltepec, had the less variation in maize yields ranging between 798 and 886kg/ha with a mean of 744kg/ha. In Class 2 lands located in Nopalucan, Ixtiyucan and Zitlaltepec, yields were variable ranging between 171 and 960kg/ha with a mean of 732kg/ha. Class 3 lands in El Seco, Soltepec, Mazapiltepec and Xicotenco, yields were extremely dissimilar and fields produced between 350 to 1108kg/ha with an average of 681kg/ha.

The eastern locations in Atenco and Aljojuca of the Llanos de San Juan have the lowest quality land Class 4 and 5. Class 4 lands yielded more homogeneously and ranged between 426 and 752kg/ha with an average 562kg/ha. The same trend was observed in fields with Class 5 lands which produced an average 482kg/ha and had ranges of 399 to 558kg/ha.

Maize yields according to municipio

Another comparative analysis of maize yields within the study region was made according to municipios. For this comparison, I calculated the average yields of the 2009 field surveyed and organized them according to their location within each municipio. This information was then compared to the historical municipio averages reported by SAGARPA for the years 2001-2008 (Figure 66). Afterwards, the 2009 anomaly percentage was estimated by dividing the 2009 average yields by the historical average and afterwards substracting 1 in order to obtain the negative decrement:

$$\text{Anomaly} = \frac{2009 \text{ average}}{\text{Historical average}} - 1$$

In general, the anomalies in maize production varied considerably between major regions during 2009 (Figure 67). At the regional level, the results show, again, that the Llanos de San Juan was the area least affected by the canícula drought registering a -69.04% anomaly from the historical average. The Llanos de San Juan area had the highest average yields (688kg/ha) for the entire study region, yet the range of production was dissimilar ranging from as low 171kg/ha to as much as 1108kg/ha. However, 97% of them recorded between 300 and 1000kg/ha (Figure 68).

The least affected were the western municipios of Mazapiltepec (-62.55%) and Nopalucan (-64.07%). This is

Agricultural Production for the Year 2009

Municipio	Region	Historical average (2001-2008)*	2009 average	Anomaly
Tecali de Herrera	Tepeaca valley	1272	99	-92.26%
Santo Tomas Hueyotlipan	Tepeaca valley	350	47	-86.72%
Tepeaca	Tepeaca valley	1961	267	-86.39%
Quecholac	Tepeaca valley	700	98	-85.96%
Cuautinchan	Tepeaca valley	840	123	-85.39%
Acajete	Puebla-Tlaxcala valley	2761	432	-84.35%
Acatzingo	Tepeaca valley	1075	170	-84.17%
Amozoc	Puebla-Tlaxcala valley	2789	481	-82.76%
Tepatlaxco	Puebla-Tlaxcala valley	2588	448	-82.69%
General Felipe Angeles	Tepeaca valley	850	165	-80.54%
Aljojuca	Llanos de San Juan	2308	462	-79.99%
Reyes de Juarez	Tepeaca valley	790	159	-79.82%
San Juan Atenco	Llanos de San Juan	2166	453	-79.08%
San Salvador Huixcolotla	Tepeaca valley	346	85	-75.29%
San Salvador el Seco	Llanos de San Juan	2213	561	-74.66%
Soltepec	Llanos de San Juan	2251	598	-73.43%
Puebla	Puebla-Tlaxcala valley	1700	480	-71.79%
Zitlaltepec	Llanos de San Juan	2214	662	-70.10%
Mixtla	Tepeaca valley	250	75	-70.01%
Cuapiaxtla	Tepeaca valley	273	96	-64.94%
Tochtepec	Tepeaca valley	187	67	-64.30%
Nopalucan	Llanos de San Juan	2206	793	-64.07%
Mazapiltepec	Llanos de San Juan	2205	826	-62.55%
Tecamachalco	Tepeaca valley	128	88	-30.92%
Average		1434.30		

*Source: www.oeidrus-df.gob.mx

FIGURE 66. HISTORICAL AVERAGE MAIZE YIELDS (KG/HA) AND THE 2009 ANOMALY ACCORDING TO MUNICIPIO.

Region	Historical average maize yields (2001-2008)	2009 average maize yields	Anomaly
Llanos de San Juan	2222	688	-69.04%
Puebla Tlaxcala valley	2459	448	-81.78%
Tepeaca valley	694	197	-71.68%

FIGURE 67. AVERAGE MAIZE YIELDS FOR 2009 AND PERCENTAGE OF ANOMALY BY REGION.

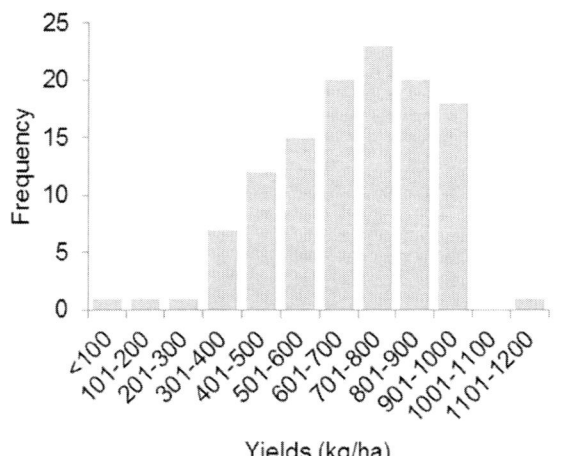

FIGURE 68. FREQUENCY OF MAIZE YIELDS DURING 2009 FROM FIELDS LOCATED IN THE LLANOS DE SAN JUAN.

concordant with the higher precipitation levels registered in this area during 2009 and the presence of better Class 1 and 2 lands. The most affected municipios were Aljojuca (-79.99%) and Atenco (-79.08%), located in the eastern portion of the Llanos de San Juan with Class 4 and 5 lands.

On the other hand, the Tepeaca valley as a whole recorded an anomaly in maize yields of -71.78%. Here, maize production by municipios was in general low and ranged from 30 to 679kg/ha with a mean of only 197kg/ha. The majority of fields (96.62%) yield less than 500kg/ha and only a few were able to reach above 600kg/ha (Figure 69). The highest anomalies were recorded in the central and southwest municipios were Class 3 and 4 lands predominate including Tecali (-92.26%), Hueyotlipan (-86.72%), and Tepeaca (-86.39%).

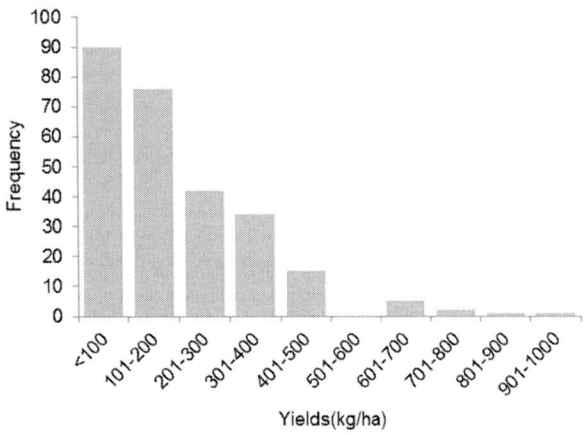

FIGURE 69. FREQUENCY OF MAIZE YIELDS FROM FIELDS LOCATED IN THE TEPEACA VALLEY DURING 2009.

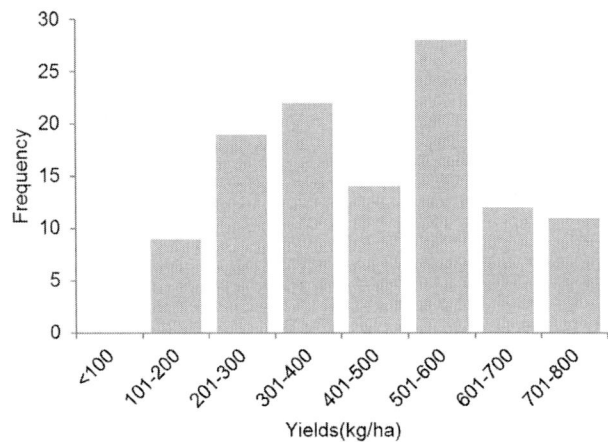

FIGURE 70. FREQUENCY OF MAIZE YIELDS FROM FIELDS LOCATED IN THE PUEBLA-TLAXCALA VALLEY DURING 2009.

Tecamachalco, Tochtepec, Cuapiaxtla, and Mixtla, located in the southeast portion of the Tepeaca valley, are an interesting case. Their anomalies were not as severe as that recorded in other municipios of the valley and ranged between -30.92% and -70.01%. One reason that can explain such a pattern is that the lands within these municipios are marginal with low precipitation regimes. As a result, this area has the worst historical maize yield averages for the valley, most of them falling below 350kg/ha. Therefore, although the 2009 canícula drought produced a decline in the rains for several months, the local maize varieties usually grow under these poor environmental conditions. In consequence, maize productivity would hardly lower more than what it usually does.

Interestingly, although the Puebla-Tlaxcala valley has better Class 1-3 lands for cultivation, it suffered the highest anomaly of -81.78%. Fields located in this area yielded between 175 to 800kg/ha (Figure 70) with a mean of 448kg/ha. Acajete was the most affected with an anomaly of -84.35%, followed by Amozoc with -82.76%. However, even though there was a severe anomaly registered the average productivity in fields of the Puebla-Tlaxcala area was twice as much as the recorded for the Tepeaca Valley. This gives some sense of the agricultural importance of this zone either to modern and prehispanic populations.

From the information presented we can conclude that in modern times the Tepeaca valley is much less productive in relation to the western Puebla-Tlaxcala valley and the northern Llanos de San Juan. Part of the explanation is based on the poorer environmental conditions, lower rainfall, and lesser quality soils found in the southern area. Nevertheless, the 2009 canícula drought produced a generalized decrease in maize yields throughout the study region. The Puebla-Tlaxcala area recorded the highest anomaly, but was able to produce a decent amount of maize even under a considerably severe disruption in precipitation. The Llanos de San Juan was the least affected zone, in particular the western portion where Class 1 lands and better precipitation levels predominate. The Llanos region had an average of 838kg/ha, a value considerably higher compared to the rest of the study region.

The area most affected by the disruption in rainfall patterns was the southwest portion of the Tepeaca valley. This zone was struck by both a lack of water or plants and the less quality soils, resulting in an anomaly of up to 92% from the historical average. Yet, either under normal or unusual climatic conditions, the southeast portion of the Tepeaca valley remains the poorest area for maize cultivation. The main restraint are the poor marginal soils of the region and the low precipitation levels, both resulting in very low yields that fall below 350kg/ha on average.

Differential sowing and harvest within the region

In general, the anomalies in maize production varied considerably between major regions during 2009 (see Figure 67). At the regional level, the results show, again, that the Llanos de San Juan was the area least affected by the canícula drought registering a -69.04% anomaly from the historical average. The Llanos de San Juan area had the highest average yields (688kg/ha) for the entire study region, yet the range of production was dissimilar ranging from as low 171kg/ha to as much as 1108kg/ha. However, 97% of fields recorded between 300 and 1000kg/ha (Figure 68).

Another aspect of regional maize production worth exploring is that of the differential harvest times within the region. Although there is a high level of idiosyncrasy among farmers regarding the timing of initial sowing, environmental factors shape the start and end the agricultural cycle. The northwestern Tepeaca region has a higher precipitation regime and deeper soils with better moisture retention than the rest of the region (see Figures 8 and 10). These factors, and the need to avoid early frosts, make farmers sow very early in the year and harvest their crops as early as possible in September (Figure 71). In high altitude areas between 2800 and 3000m, it is possible to cultivate maize using long developing varieties that

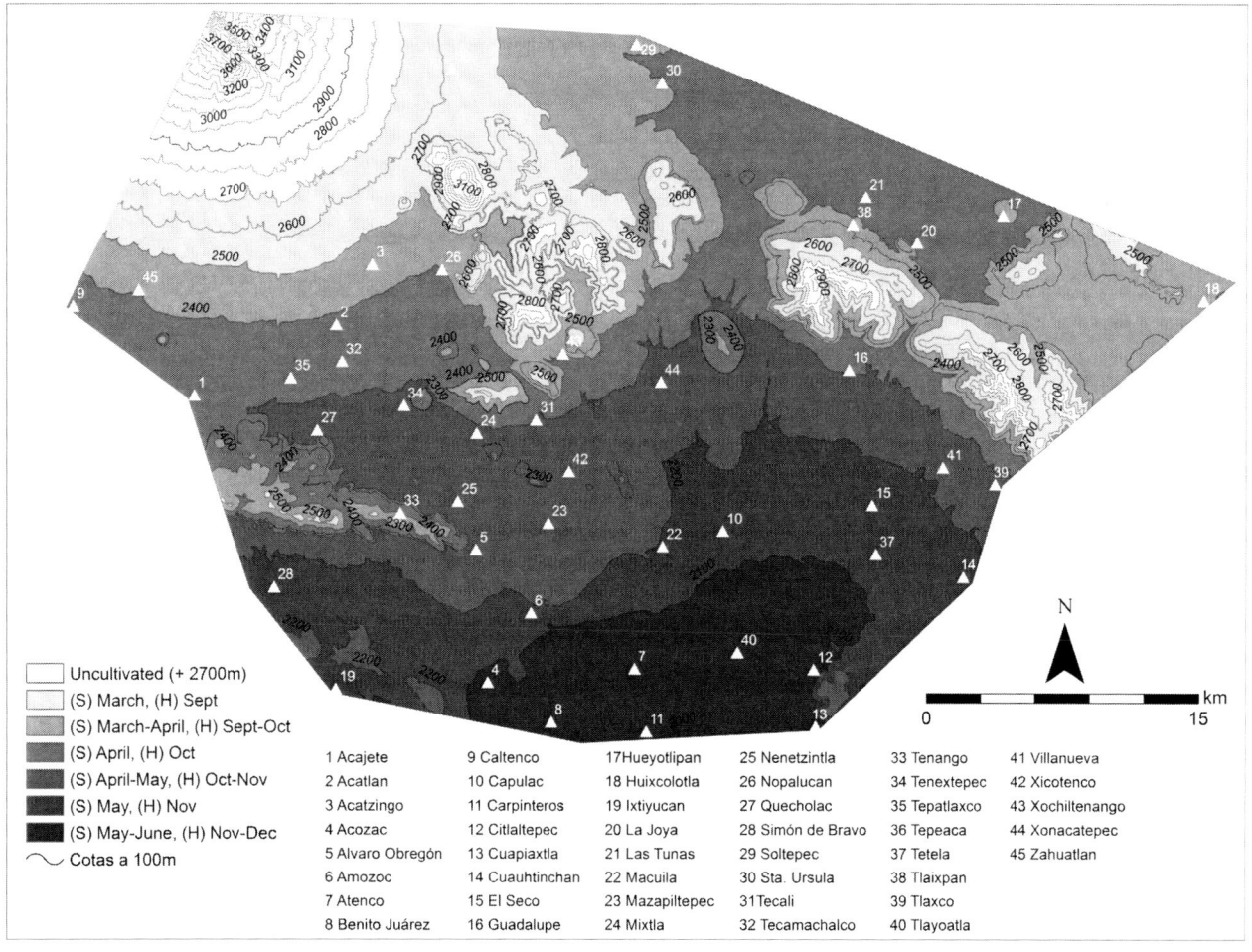

FIGURE 71. APPROXIMATE DISTRIBUTION OF THE SOWING (S) AND HARVEST (H) MONTHS FOR SIX-MONTH MAIZE VARIETIES IN THE TEPEACA REGION.

are sown as early as March and are harvested as late as December and January (Lara, *et al.* 2002).

In the southern area around Tecamachalco, Quecholac, Mixtla and Tecali, soils have low moisture retention and a high evo-transpiration rate. There, cultivation takes place around May and July to take advantage of the higher precipitation levels. These farmers usually harvest between November and December.

In the central portion of the valley around Tepeaca, Acatzingo and Macuilá, the agricultural cycle usually starts in April and May with harvest taking place between October and November. In the southwest around Cuauhtinchan, the cycle extends from May to late October and November.

In addition, maize landraces have different growing-harvest cycles. Some are very precocious and grow rapidly while others take longer period to arrive at maturity. Gil and his colleagues (2004) registered growth cycles for 2514 samples collected in 15 micro-regions in Puebla, including areas within the Tepeaca region. They found a difference of 105 days between the most precocious landrace (Ayotoxco) located in low altitudes (140m) and the long developing ones (La Malinche) in high altitudes (2760m).

Both individual decision-making and the use of a variety of different local maize landraces produce a mosaic of harvests taking place at different times. This results in a constant supply of maize to local warehouses and marketplaces from as early as September to as late as January. Such a pattern could have been very useful for the prehispanic and Early Colonial Tepeaca populations. It could have signified a constant supply of food provisions throughout most part of the year.

In this chapter I have analyzed and outlined the key variables that influenced differences in agricultural production between single households at the regional scale. I have done so by conducting a regional survey of 49 fields from 24 households within the study area. This small sample was later expanded to 500 fields in order to have a better picture of the effects of the 2009 canícula drought on regional maize yields. The results show that severe regional climatic perturbations such as the canícula drought can have important repercussions on agricultural systems and food production at the household and regional levels. However, the effects of the 2009 dry canícula varied throughout the study region. The drought was most severe in the central and southwest portion of the Tepeaca Valley in areas like Tecali, Cuauhtinchan and Tepeaca, and in Amozoc and Acajete on the western Puebla-Tlaxcala valley. The least affected areas were the northwest portion of the Llanos de San Juan.

The 2009 ethnographic data revealed that buffering strategies employed by local smallholders against severe climatic disruption (e.g., field dispersion and differential sowing/harvest patterns) were not sufficient to mitigate the profound negative effects of the canícula. The data obtained from the survey of 500 fields was extremely important. Mexico's SAGARPA's often records annual maize production levels of zero as a result of environmental affectations on cultivars (see Figure 57). As it was demonstrated in this chapter, this does not correspond with reality. Modern farmers may report total crop failure to governmental agencies in order to obtain emergency funds. However, as this chapter has shown, every area did suffer some degree of crop failure. Yet, the 2009 survey found no case in which a field had total crop failure. At least some portion of the crop stands will survive and be able to produce maize.

In Chapter 8, I use the data derived from the ethnographic study of maize production and environmental variability to model past buffering strategies for food acquisition at both the subsistence agriculture and political institutional agricultural levels.

Chapter 7
From Prehispanic *Macehualli* to Colonial *Terrazgueros*

In Chapter 4, I briefly described the characteristics of the landless sector of Early Colonial Tepeaca known as the *terrazgueros*. In this chapter I describe in greater detail their role as the main supporters for the political institutions of the *altepetl*. The social and tributary relationships between *macehualli* and noble elite underwent profound changes during the Colonial period. Examining these relationships is important considering that such changes were reflected in the Tepeaca historical accounts.

The term *terrazguero* was employed by the Spanish to identify a particular group of people that rented land from nobles. Hildeberto Martínez (1984b) has shown that the *terrazgueros* of Colonial Tepeaca were landless peasants that comprised a large proportion of the population. *Terrazgueros* lived on a noble's estate, paying him tribute in service, produce, and labor for the right to cultivate his lands. An indigenous Nahuatl term may have been used to identify them, but unfortunately, the Tepeaca colonial documents lack such vocabulary. However, in the Basin of Mexico and in Tlaxcala, Colonial documents commonly speak of a landless sector known as *tlalmaitl* or *mayeque* whose social and economic organization was described amply by 16th century AD Jurist Alonso de Zorita (1942). Carrasco (1989: 126-127) discusses the meaning of this Nahuatl word and concludes that in a metaphorical sense means a "working man" (*trabajador*). However, in his dictionary Alonso de Molina (2008 [1571]) includes word *tlalmaitl* and translates it as a "*labrador o gañan*", meaning a farmer and literally the "hands of the land". *Mayeque* may be a Nahuatl word incorporated into Spanish that derived from the plural of *tlalmaitl*, which would be *tlalmayeque* or simply *mayeque*. The *mayeque* share a similar landless condition and patronage relationship with local elite as did the *terrazgueros* from Tepeaca. The *mayeque* term is used principally in colonial texts and could also be a Nahuatl word incorporated into Spanish. *Terrazgueros* and *mayeque* are likely the same people, or at least a very similar social category that existed both in the Basin of Mexico and in the Puebla-Tlaxcala region, but expressed in different regional terms.

In my view, there is little historical evidence in Tepeaca to support the strong separation of a landless renting sector during prehispanic times as it occurred in Colonial times. Prior to the Colonial regime, the system was probably based on tributary demands over dominated and conquered populations, as it is described in historical documents from the Puebla-Tlaxcala regions(Kirchhoff *et al.* 1976; Martínez 1984a, 1994a; Molina 1985 [1580] ; Muñoz Camargo 1998 [1580] ; Prem 1978; Reyes 1988b).

The Tepeaca region historical texts pertaining to events prior to the Spanish arrival consistently employ the term *macehualli*, while neither the term *terrazguero* nor any equivalent in Nahuatl can be found. The term *macehualli* possibly incorporated many forms of tributary people, including those with patrimonial lands, landless people, and individuals who sought additional areas for cultivation. Even though the legal documents from Tepeaca indicate that the nobles possessed large quantities of land, it's clear that some sectors of the *macehualli* population had access to patrimonial plots. For example, the testament of the tlahtoani Diego Ceinos establishes that:

> …and the many lands the they enjoyed and took advantage of because to each one of the mentioned renters five and six and seven and eight pieces of land were given of six braza wide and one hundred long…although some of them had other [lands] of their patrimony in which they did not recognize any person [as possessing them] and if the aforementioned indian renters would not understand and see the benefit and utility of tilling and taking benefit of them by way of rent it would be better that they leave them and be content with the patrimonial ones they had… (Carrasco 1973: 32) (my translation).

The term *macehualli* was generally employed to refer to non-noble people. It was also used to name individuals of noble origin and *calpulleque* that had been enforced as tributaries by other nobles.[iii] Therefore, *macehualli* appears to have been a general word employed for a tributary person or a vassal as Alonso de Molina (2008 [1571]) translated it. This seems to be a particular element in the prehispanic social organization of the Cuauhtinchan-Tepeaca region.

As I explain below, the conceptualization of a *terrazguero* sector in Tepeaca may have appeared as a result of the social and economic reconfiguration that occurred with the Spanish Conquest. The term *terrazguero* was widely used in legal disputes, land transactions, patrimonial relations and testaments of indigenous nobles from

---

[iii] "Y decimos, declaramos, de este modo sabemos, los tolteca chichimeca que ahora entablan pleito, los que se nombran *calpulleque*, que dicen que es tierra de ellos, en verdad es tierra que les pertenece. Les fue quitada [su tierra] y fueron convertidos en macehualli por medio de opresión. Y los dos tlahtoani Couatecatl y el Tezcacoatl en nada oprimieron a la gente, a nadie convirtieron en macehualli suyos, ninguna tierra perteneciente a otros tomaron; sino que algunas de sus tierras les fueron quitadas y a otros teuhctli [también] les fue quitada su tierra. Solamente fueron dos los que tomaron tierras ajenas y convirtieron a la gente en macehualli suyos, son Totomochtzin y Cuitlauatzin y Tezcocollo, el tercero" (Reyes 1988b: 86-87).

Tepeaca and other contemporaneous settlements. In 1560, the indigenous cabildo of Huexotzingo explicitly stated that *macehualli/terrazguero* of nobles were those individuals that lived in their patrimonial lands, pay them a terrazgo or rent, and did not have lands of their own (Prem 1978: 58).[iv] *Terrazgueros* were the main economic support for the declining indigenous political institutions during the Early Colonial period. They were one of the last resources over which nobles could maintain political power. As a strategy, nobles claimed legal rights over vast land resources using the new Spanish legal system as a catapult. With it, they claimed a tributary dominance over *macehualli/terrazguero* that were settled on their lands.

In the rest of this chapter I explore the main social and economic differences between the prehispanic *macehualli* and the Colonial *terrazguero*. The hierarchical role of the *macehualli* was influenced by the numerous conquest episodes that characterized the Postclassic period in central Mexico. I examine historical data that provides insights on changes in the land tenure organization during the four centuries prior to the Spanish Conquest. Conflicts between rival *altepemeh* affected the social and political relations of indigenous populations and augmented the sharp polarization in access to land.

The prehispanic *macehualli*

The role of the *macehualli* populations from the Tepeaca region and adjacent areas has been amply studied (Olivera 1973, 1978), but two sets of historical information provide insightful data regarding land tenure changes associated with conflict and conquest events: (1) the *Historia Tolteca-Chichimeca* and (2) the collections of historical documents from Tepeaca and Cuauhtinchan. These documents show that during the Colonial period access to land in the Tepeaca region was highly polarized (See Chapter 4). The nobility controlled a large proportion of the land while non-nobles were mainly landless. Unfortunately, most studies conducted in Tepeaca and nearby regions have analyzed the shifts in power and land control mainly from a political economy perspective. We know much less about the peasant households that supported the institutional apparatus because they are poorly represented in these documents.

The sharp difference in wealth and access to land between the *macehualli* and the Tepeaca noble elite in the Early Colonial period was partly the result of various conquest events of the Postclassic period. Some of the major changes in the political and territorial reorganization in Tepeaca may have occurred as early as the 12th century AD. Chichimec invasions and post-conquest conflicts between rivalry *altepemeh* were key factors for the appearance of a highly polarized access to land between nobles and commoners. Unfortunately, most of the historical data available pertains to events written from the perspective of the Chichimec elite. Although the native history is biased, some data can be extracted to inform us about the economic importance of the *macehualtin* populations. Of particular relevance to elites were land and labor because they supplied the staple products that supported the political institutions and financed stratification.

Chichimec conquests in the Cuauhtinchan-Tepeaca region

The Postclassic period was characterized by several Chichimec migrations and conquests in the central Puebla and Tlaxcala regions. The Chichimec as a group consisted of a number of highly belligerent societies that sought to overthrow the local ruling groups and take over regional power, political territories, and tributary populations. The *Historia Tolteca-Chichimeca* tells that the first Chichimec groups invaded the Cuauhtinchan and Tepeaca regions around 1170 A.D and were known as the moquiuixca-cuauhtinchatlaca. The moquiuixca expanded toward the eastern territories of today's central State of Puebla resulted in the subjugation of several ethnic groups. They settled on the southern side of the *Sierra de Tepeaca* mountain range, founding the Cuauhtinchan *altepetl*. Their leader, named Teuhctlecozauhqui, became the first Chichimec tlahtoani in the region and his noble linage ruled for more than 200 years. A few years after the arrival of the moquiuixca, the colhuaque (AD 1178) and the tepeyacatlaca (AD 1182) also entered the region and settled at the east edge of the Amozoc-Tepeaca mountain chain. These two groups became key allies in the expansion of the Cuauhtinchan domains and played an important military role in the subjugation of settlements in the northern and eastern Tepeaca valley.

The Chichimec conquests were aimed at taking political control of territories. Conquered populations were deprived of their lands and the defeated elites were either killed or expulsed from the region. The rest of the populations were taken as *macehualli* tributaries. The defeated communities were composed of varied ethnic groups, sometimes with different forms of social organization. Yet the Chichimec consigned all the non-noble people, and possibly some noble individuals, to the *macehualli* sector. This emphasized their tributary role to the conquerors and gave all conquered people equal social status despite their clear internal social and ethnic differences (Olivera 1978: 171). This resulted in a twofold social division consisting of the conquering noble ruling sector and all others grouped together as *macehualli* tributaries. Some local nobles may have been retained by the Chichimec as leaders of their respective communities or linked as allies through strategic marriages, as was the trend during the Postclassic (Berdan *et al.* 1996). It is probable that the division between non-nobles and elite was already an important component of societies before the arrival of Chichimec conquerors. For example, we know from the *Historia Tolteca-Chichimeca*, and the writings of 16th century AD chronicler Diego Muñoz Camargo, that the famous ruling olmeca-xicallanca of Cacaxtla and Cholula distinguished themselves from the

---

[iv] " ...son macehuales terrazgueros de principales de la dicha provincia que están en tierras de sus patrimonios y que les pagan terrazgo de sus tierras y no tienen tierras propias suyas..."

dominated tributary populations, such as the Toltec groups in Cholula (Jiménez 1995). Unfortunately, we lack more information on social hierarchies for ethnic groups prior to the Chichimec invasions.

It is difficult to determine from the historical accounts if the Chichimec invaders brought with them non-noble individuals as part of their military and logistical support. We know that an army of men was provided to the cuauhtinchatlaca moquiuixca by the newly enthroned Toltec elite from Cholula. With all probability, some *macehualtin* formed part of the conquest expansion of Teuhctlecozauhqui. Some commoners probably achieved a higher hierarchical status and access to land resources by performing extraordinary military accomplishments. This was a basic trend behind the Chichimec conquests.

Between AD 1234 and 1246 there were various migrations of Toltec-Chichimec *calpulli* from Cholula to Cuauhtinchan. The *calpulli*, as discussed in Chapter 3, was a distinctive corporative institution where several households were affiliated based on kinship ties, shared patrimonial lands, and kept control of various internal issues. The *calpulli* migrations to Cuauhtinchan were motivated by a large famine that struck Cholula after its war against Huexotzingo and the Acolhua of the Basin of Mexico. The *Historia Tolteca-Chichimeca* indicates that around 20 *calpulleque* resettled in the Cuauhtinchan *altepetl* territories. No specific numbers of households or persons are provided, but it is possible that they were high.

The Toltec *calpulleque* from Cholula were allied with the Cuauhtinchan Teuhctlecozauhqui lineage through strategic marriages. According a 1553 manuscript from Cuauhtinchan these *calpulleque* were granted asylum and were provided with patrimonial lands; they were also elevated to the rank of *teuhctli* and *pilli* and exempted from paying tribute to the local noble elite, thus avoiding the tributary status of *macehualli* (Reyes 1988b: 86-87). Furthermore, they were allowed autonomy on internal issues, although some tribute arrangements may have been arranged with the local tlahtoani ruler. Thus, the *calpulleque* from Cuauhtinchan appear with a different social status and organization structural group that were related politically and biologically to the Cuauhtinchan Chichimec elite. They were not considered simple commoners and did not from part of the *macehualli* sector. *Calpulleque* acted as a relatively independent social and economic institution coexisting alongside the noble ruling elite and the *macehualli* population. The *calpulli* was not an institution present in the Cuauhtinchan-Tepeaca area prior to the 13th century AD.

Late in the 14th century AD, a series of internal conflicts were initiated at the core of the Cuauhtinchan *altepetl*. By this time, several ethnic groups (*parcialidades*) lived within its boundaries generating political tensions. In a crucial episode, disagreements between two groups of Chichimec nobles ended with the fission of one elite group and its resettlement in nearby Tecalco (today Tecali).

Seeking to end the disputes, the Tecalco nobles solicited the intervention of the Mexica-Tlatelolco of the Basin of Mexico, a powerful group in that region. The Tlatelolca decided to interfere militarily and in 1398 A.D they conquered Cuauhtinchan, ending the rule of the Chichimec linage of Teuhctlecozauhqui. The Tlatelolca gave political control of the *altepetl* to the Mixtec Pinome. This change had negative effects at the regional level and set off a strong rivalry with other nearby settlements, especially with Tepeaca.

The Tlatelolca conquest of Cuauhtinchan also generated another *calpulli* migration from Cholula. The 1553 legal manuscript from Cuauhtinchan (Reyes 1988b: 82) mentions that with the death of the Teuhctlecozauhqui lineage rulers, another wave of *calpulleque* immigrated and took possession of some abandoned lands in the area between Cuauhtinchan and Tepeaca.

While Cuauhtinchan was ruled by the Mixtec Pinome, a Chichimec majority ruled in Tepeaca. Tepeaca's power grew considerably in the 15th century AD. In AD 1457 Tepeaca initiated a war against Cuauhtinchan seeking its independence from the center. The war lasted two years and finally ended in AD 1459 with the defeat of the Pinome elite and their expulsion to the southern region of Matlactzingo. During the next two years, no group had political control over the unguarded Cuauhtinchan territories. In the following seven years, Tepeaca invaded these lands and started to collect tribute from the local *macehualtin* communities. The historical texts omit any information on *macehualli* participation in the military events because they were, in part, the center of the disputes. However, the *macehualtin* were a social class that would have participated in all the disputes alongside the nobles as warriors as part of their tribute obligations. After one group defeated the other, *macehualtin* tribute obligations were simply transferred to the winning side. Local land tenure arrangements probably remained unchanged after external conquest.

Defeated and in exile, the Pinome solicited military support from the Mexica Empire in order to recover their lost political power. The Mexica clearly understood the economic benefits that the Tepeaca region offered due to its strategic location along the commercial trading routes leading towards the southern and eastern lowlands (Martínez 1994b). As a result, the Mexica conquered Tepeaca in AD 1466. This conquest produced a major territorial rearrangement. All the Cuauhtinchan *altepetl* was partitioned into five new territorial entities: Cuauhtinchan, Tepeaca, Tecamachalco, Tecalco and Quecholac. This arrangement lasted up into Colonial times (Yoneda 1991).

With the Tepeaca Conquest, the Mixtec Pinome regained their power over Cuauhtinchan. Yet, their territorial domain was considerably diminished. On the other hand, Tepeaca obtained its own territory and was ordered to install an important regional marketplace in the capital town. Tepeaca also became a regional tribute collection center for the Mexica which gave the Tepeaca *altepetl* a high level of regional economic and political power (Martínez 1994b).

The Mexica conquest had strong repercussions on local land tenure arrangements. On the one hand, there was a massive resettlement of populations into Tepeaca to take advantage of the newly established marketplace. This immigration of population included several *calpulleque* that moved to Tepeaca from Cuauhtinchan. On the other hand, political alliances of some Tepeaca elite with Mexica lords promoted a new series of internal conflicts between some of Tepeaca's *teuhctli* and with the *calpulleque*. Around AD 1492, a massive dispossession of *calpulleque* and *teccalleque* lands was undertaken by two Chichimec lords, Totomochtzin and Cuitlauatzin. These lords robbed lands through force and assassinations. Several *calpulleque* were stripped of their territories and rights and were converted into *macehualli* tributaries (Reyes 1988b: 99). The appropriation of *calpulli* and *teccalli* lands augmented the already polarized social and economic inequality and augmented the number of *macehualli* people. This situation prevailed until the Spanish arrival.

Colonial *macehualli* and *terrazguero*

When the Spaniards conquered Tepeaca in 1520, they immediately established a new socio-political order. The effects of the Spanish Conquest include the reorganization of political relations, the breaking up of military and religious institutions, and a large demographic collapse due to the introduction of new diseases for which indigenous populations had not been exposed. Enslavement, war, abusive labor demands, and excessive tributary payments created a demographic decline of 50% to 90% of the indigenous populations in central Mexico (McCaa 1995; Trautmann 1997). The effects of the epidemics are best exemplified by Juan Bautista Pomar (1941: 50) a colonial chronicler from Tezcoco:

> ...and it is not found that his parents or ancestors gave news of ever being such pestilence nor mortality, as after their conversion it has happened to them, so large and cruel that it is affirmed it was consumed for every ten parts the nine of the people that there was... because they affirmed that it was countless the people that there were, and it seems clear it was so for the vast quantity of land they worked and farmed, that today appears furrowed generally everywhere, the majority deserted and weeded, and with three general pestilences that has had since they were won [conquered] they have consumed and decayed...[my translation from Spanish].

In Tepeaca the Conquest weakened the *teccalli* economic system. In particular, it affected the economic relationship between nobles and *macehualli*. Nobles no longer had political or military power over commoners who were now under the patronage of Spanish cleric missionaries and the new encomienda system. Indigenous nobles were forced to develop new ways to perpetuate the *teccalli* system. One strategy was to maintain legal control over large land estates and with it the landless *macehualli/terrazguero* sector. The organizational confusion generated by the Conquest allowed some nobles to appropriate lands that had belonged previously to commoner populations (Prem 1978). Nobles sought to demonstrate that they owned all land and that such right had been obtained by way of conquest prior to Spanish arrival:

> The first thing I say, I declare, is that here in Cuauhtinchan, in Tecalco, in Tepeyacac, in Tecamachalco and in Quecholac, the *calpulli* do not possess lands. The manner in which the lands are arranged [is] that the fields are not in the properties of the calpulleque. Only the tlahtoani had lands and in them they serve as tlahtoani, in them they favored the macehualli, they help them (Reyes 1998b: 83) (my translation).

Controlling the land allowed nobles to demand compensation for their use. On the contrary, *macehualli* argued that nobles had maliciously appropriated lands that were rightfully theirs during past decades. It is possible that some of these disputes were carried out by the *calpulleque* which as group had been strongly affected by such land usurpations.

The new Spanish order also promoted the idea of massive dispossession of land by nobles prior to the Spanish arrival. Sixteenth century AD writers like Martín Cortés and Vasco de Puga spread the image of the indigenous lords as greedy tyrants who took advantage of their social position by appropriating lands from *macehualli* (Martínez 1994a: 83). However, this slander was directed toward indigenous lords because the Spanish sought to acquire the rights over the indigenous noble lands. By attacking the system of prehispanic land tenure, the new Colonial ruling apparatus had a basis for reassigning indigenous lands to Hispanic colonizers. Some noble lands were uncultivated and in fallow leading the Spanish to considered them as 'no man's lands' (Martínez 1994a). The argument was that, if the Crown did not expropriate these uncultivated areas and distributed them among the new Spanish colonizers, the indigenous nobles would appropriate them. In many ways, land disputes led to a more exacerbated division between nobles and *macehualli* than had previously existed.

In prehispanic times, war and conquest were geared towards taking political control of a region. The victors generally retained native populations and incorporated them into the economic system as tributaries. As a result *macehualli* populations shared no affinity with the ruling overlords. This view was so embedded into indigenous perceptions, that after the Spanish defeated the Tepeaca nobles in 1520 and took political control over the populations, the *macehualli* sector considered that tribute payments should go directly to the Spanish Crown; they explicitly refused to continue as tributaries of the indigenous nobles (Martínez 1984a: 466-467).

In Tepeaca, most legal disputes during the colonial period ended in favor of the local nobles despite *macehualli* reclamations. This led to a new arrangement between nobles and landless peasants. *Macehualli* were allowed to settle and take advantage of the noble's agricultural fields in return for tribute payments in both service and in products. It is precisely during the first half of the 16th century AD and until AD 1560 when these *macehualli* renters, termed *terrazgueros* by the Spanish, became particularly prominent in legal documents.[v] *Terrazgueros* consistently form part of legal disputes and documents dealing with property transfers, inheritance, and the distribution of working lands.

The demands for the payment of rent from landless peasants became the most efficient way for indigenous lords to perpetuate the *teccalli* system for most of the 16th century AD. For nobles, the focus of exploitation changed from a military and ideological tribute demand on *macehualli* populations during the Postclassic period, to one based on the payment of a rent during the Colonial period. Those landless *macehualli* that lived on the landed estates of a noble, which were probably relatively fewer in relation to the number of ordinary *macehualli* dispersed in various populations, became known generically as *terrazgueros* by the Spanish and probably *tlalmaitl* or *mayeque* in other areas by the local lords.

By 1560 specific political actions of the Spanish Crown indicate that its intention was to dissolve the noble *teccalli* institutions. Censuses were undertaken midway the 16th century AD in order to ascertain the number of potential tributaries for the Spanish Crown and those considered tributaries of noble lords. Traditionally, landless individuals were exempt of tribute obligations to the Spanish because they were considered as tributaries only to a noble lord. For this reason, cadastral registers, such as the Matrícula de Huexotzingo, clearly differentiate between *macehualli/ terrazguero* (identified with a red dot) and *macehualli* with patrimonial lands (Flores-Márquez *et al.* 2006). Several nobles were persuaded to donate lands to the landless *macehualli*, while others were simply deprived of them and were absorbed by the Encomienda system or relocated directly as vassals to the Spanish Crown (Carrasco 1966: 146-148; Reyes 1988a; Zorita 1942: 132-133). In particular, the foundation of the Spanish city of Puebla de los Angeles, some 30km west of Tepeaca, relied on indigenous labor force from the Tepeaca valley and adjacent regions. This probably detached some proportion of *terrazgueros*, and perhaps some *calpulli* sectors, from their tributary obligations towards nobles. By the start of the 17th century AD, both Spanish demands for labor and the decline of *terrazgueros* due to disease had eroded the indigenous economic infrastructure based on the land rent. This set the stage for a massive selling of land on part of the indigenous nobility and the decline of most indigenous political institutions of Tepeaca (Perkins 2007).

In this chapter, I described in detail the characteristics of the landless sector of Early Colonial Tepeaca known as *terrazgueros*. I also examined the role of tributary *macehualli* populations and their social and economic relationship with noble elite, centering on the profound changes they underwent from the Postclassic to the Colonial period. This information is useful in order to better understand the role of commoner populations as economic supporters of political institutions in ancient Tepeaca. Such economic relationship, including their role as agricultural and wealth item producers, is further discussed in Chapter 8.

---

[v] '…hago presentación de esta memoria y pintura por la qual consta de todas las tierras que el dicho mi menor [Esteban de Mendoza] tiene y posee los maceguales renteros…' (Archivo General de la Nación Tierras, Vol. 2782, exp. 40, f.11r).
'…[Dionisio de Mendoza] tiene sus sementeras que le siembran sus terrazgueros; las quales siembran los principales sus parientes descendentes de su casa y señorío, las quales siembran sus maceguales…' (Archivo General de la Nación Tierras, vol. 2782, exp. 40, f. 7v-8r).

Chapter 8
Agricultural Productivity and Tribute in 16th Century AD Tepeaca

In this chapter, I examine the role of the *macehualli/ terrazguero* as the agricultural supporters of Tepeaca political institutions. I do this by estimating the amount of land allotted to landless *terrazgueros* and the level of crop production of local agricultural systems. With this information, we can estimate the quantity of food production for both household self-sufficiency and tribute payments. This also allows a direct comparison between food production at the domestic level and the demand of political institutions on the commoner household's food production base.

Land allotment and agricultural tribute in Early Colonial Tepeaca

As I have shown in the previous chapter, control over the *macehualli/terrazguero* sector was one of the most important economic supports for indigenous political institutions. In particular, agricultural production was a major means for financing the existing system of stratification. Despite abrupt changes associated with the Spanish conquest, agricultural production continued to be an important way of generating surplus that allowed nobles to persist as elite within colonial society. Yet, it was the peasant commoner household that shouldered the burden of supporting the ruling apparatus. For them, agricultural tribute was an imposition that could, in some degree, intervene with their sustenance base. However, it is difficult to determine to what extent political institutions interfered with staple goods production and how it affected the survival and economic well-being of the *terrazguero* household. Therefore, in this chapter I examine the direct or indirect effects of political institutions on the subsistence base of commoner households at Tepeaca.

Local historical texts from the Early Colonial period provide useful data regarding land allocations between the ruling noble sector and the commoner peasant majority. Additionally, they provide insights about the environmental surroundings and the type of agriculture systems prevalent in the region. Unfortunately, in many cases, most of the material pertains to the upper class nobility. Little information is available about the commoner sectors. Yet, there is sufficient data to permit a few relevant inferences about *macehualli* behavior.

In Tepeaca we are provided with information about the amount of land that households worked annually to sustain themselves as well as the amount of land worked for tributary purposes. As will be shown later, fields for these two purposes were well defined in terms size and allocation. This may be the result of elite awareness not to overexploit the tribute as much as it was to structure the production of wealth goods. Wealth items played an important role in the construction of socio-political relations because they were used in the creation and maintenance of strategic political alliances banquets, ceremonies, and public rituals (Brumfiel and Earle 1987; D'Altroy and Earle 1985; Hayden 2001). For Tepeaca, these goods were generated by performing an array of economic activities in addition to agriculture, probably through multi-crafting and intermittent craft production (Hirth 2009).

Land tenure in Early Colonial Tepeaca

The *teccalli* played a crucial role in the regional economy. It was the principal means for the nobles to access lands and mobilize human resources. This system promoted the highly polarized system of land tenure seen in the Late Postclassic and Early Colonial periods. Nobles controlled vast extensions of patrimonial lands and large numbers of *macehualli/terrazguero* tributaries within their domains. *Terrazgueros* were renters and were responsible for generating the necessary agricultural food and represented the sustenance base of the community. Most of the noble's lands were worked by *macehualli/terrazguero* populations. However, other sectors would also solicit rental land including merchants and bricklayers (Martínez 1984a: 447-448, 454-455). It's not clear why some individuals needed additional lands. Population pressure or scarcity of food resources may have been factors that account for this problem. This may be the case if we consider the high number of inhabitants in the area prior to the conquest and the strong political conflicts between rivalry *altepemeh* over large uninhabited areas in the boundary zones. Yet, Luis Reyes (1988a: 114) has argued that it was the desire of the local nobles to control the *altepetl* territories that best explains the land tenure relationships at the time of the conquest.

In Tepeaca, terrazgo or land rent functioned as an important component of the local political economy. The rent arrangement stipulated that the *terrazguero* would pay the rent with agricultural work, personal services, communal labor (*tlacalaquilli*) and craft production (Martínez 1984b: 97-103). Agricultural labor was a crucial element of this relationship because food production functioned as an important support for political institutions. Commonly, noble lords allotted 5 to 9 standardized plots to each commoner household, all measuring 6 by 100 *braza* (fathoms)[vi] in size, that were used for their support. One of

---
[vi] "...porque a cada uno de los dichos renteros se les dava por sus partes a cinco y a seis y a siete y a ocho suertes de tierra de seys brazas de ancho y ciento de largo, los cuales atento a el dicho provecho procuraban labrar

these specifically was set aside and worked to produce the crops used as the payment of rent.

Another form of agricultural tribute is recorded in the *Relación de Tepeaca* (Molina 1985 [1580]). Written in 1580, this document provides information on the prevalent tributary order. It is said that 400 vassals were obliged to work one field for their lords that was 400 x 400 *braza* in size; a *braza* measured from the foot to the vertically outstretched hand. The term vassal, according to the European system, refers to a feud that the individual was obligated to pay. In this contract, the rulers and nobles of the Middle Ages granted lands in usufruct, forcing those who received them to remain loyal to the lord, provide him with military service and attend the political and judicial meetings that he summoned. The fact that this document discusses tributary relationships, and that the tributaries are named as vassals, suggests that the relationship between these individuals and the local ruler was distinct from the one operating in the *terrazguero*/*macehualli* sector. These vassals are most probably *macehualli* with a similar status to those populations or groups that hold patrimonial lands described by Alonso de Zorita (1942) and Martín Cortés (1865), but were under the political control of the local elite. However, their tribute obligations were large and involved the payment of wealth and utilitarian products, as well as personal services. This arrangement established the right of peasant households to cultivate the land and, considering that the ruling noble class controlled all the land, it became a prominent way for elite to generate resources for their use. Yet, the work of Martín Cortés indicates that *macehualli* could not be dispossessed of their plots as long as they continued to pay their corresponding tribute.

It is probable, therefore, that two different agricultural tributary forms coexisted in Tepeaca. These were (1) a *macehualli*/*terrazguero* sector that tilled an area of 600 squared *braza* for the payment of rent, and (2) a sector made up of *macehualli* with patrimonial lands that paid tribute to the main ruler by working special fields of 400 squared *braza* in size for their lord.

Types of length measures in Tepeaca

Determining the actual area worked by the *macehualli* sector in terms of modern metric units is not a simple task. Fortunately, a variety of local historical texts pertaining to land allotments are available that use ancient indigenous unit measurements (Carrasco 1963, 1969; Martínez 1984a; Reyes 1988b), as well as comparative materials from adjacent regions (e.g. Castillo 1972; Cline 1986; González de Cossío 1952; Matías 1984; Sullivan 1987; Williams and Jorge y Jorge 2008; Williams and Harvey 1997). The problem researchers face, however, is a lack of concrete equivalences of pre-Columbian land unit measurements with our modern metric system. Spanish terms such as the *braza* appear to have included various forms of measuring, some contrastingly different (Gibson 2003: 263-264). Due to this ambivalence, it may be as Lockhart (1992: 143-144) has pointed out, each *altepetl* had its own set of measures. Nonetheless, these measures certainly were used in some standardized way at the local level to ensure the constant distribution and cadastral registry of fields allocated to *macehualli* households. In the following paragraphs, I provide an estimation of equivalences between modern and pre-Columbian systems. Occasionally, I also present direct quotes translated from local historical documents, all of which are written in Spanish.

The braza and the nehuitzantli

The three most important length measures employed to delimit plots for *terrazgo* use and tribute were the *braza* (*cemmatl*), the *braza* from the foot to the hand (*nehuitzantli*), and the rod to measure fields (*tlalquahuitl*). Other forms of measurement are also present, but were not included in this study, such as the "*caballería*", the "*fanega de sembradura*" and the "*mecate*". These are usually associated with large scale buying and selling transactions, land assignments, transfers, and the rent of large estates between nobles (e.g., Martínez 1984b: 191-206). The *braza* is the most frequently used length measurement found in the Tepeaca historical texts. Although it is a Castilian word, it probably is a translation of the Nahuatl *cemmatl*, literally meaning 'one hand' or 'one arm', which would be a horizontal fathom. The context in which this measure is presented suggests that the field's perimeter, or its area, were calculated using the indigenous system. Scribes simply translated the word to Spanish. In support of this argument is the fact that (1) the majority of lands recorded in disputes or transactions belonged to local indigenous individuals, and (2) the ordinary length measure used by Spaniards was the *vara Castellana*, also known as the *vara de Burgos*, declared as the standard in the American colonies in 1568 (Glockner 1991; Ruiz 1993).

The value of the *cemmatl* is obtained by computing the distance from one hand to the other with the arms extended horizontally (Figure 72). The Spaniards indicate that the value of the *cemmatl* was equal to two *varas* de Burgos (each one 0.8359m), or around 1.67m. This length concords well with the average body proportions of indigenous populations, which generally fell below 1.65m in height (Jaén *et al.* 1976; Lagunas and López 2004). Ergonomically, the *cemmatl* of a person is the same as the value obtained from his height. Therefore, the equivalence fixed by the Spanish seems to be correct, and we can establish the average value of the *cemmatl* as 1.67m in length.

In contrast, a second type of indigenous measurement termed the *nehuitzantli* was what the Spaniards called the *braza* from 'the foot to the hand'. The word is composed of the *nehuatl* (me) and *uizantli* (pointed, sharp) and

---

y cultivar con cargos de la dicha renta…" (Archivo General de la Nación Tierras, Vol. 60, Exp. 2, F 75r-87v).

"…todo esto me davan los dichos yndios por razón de bivir y estar en mis tierras, a los quales tengo repartidos a siete y a ocho y a nueve suertes" (Archivo General de la Nación Tierras, Vol. 60, Exp. 2, F44r-55v).

Date	Text	Source
1571	"Cada yndio le hazía seis brazas de sementera de ancho e ciento de largo."	AGI Audiencia de México, legajo 94, Exp. 4
1571	"Quatro yndios mercaderes…le avían de hazer tres brazas de sementera de ancho y ochenta de largo."	AGI Audiencia de México, legajo 94, Exp. 5
1571	"Le sembraba cada indio seis brazas de sementera de ancho, e ciento de largo."	AGN Tierras, vol. 2676, Exp 11
1571	"Quatro indios albañiles…le hazían dos brazas de sementera de ancho e ciento de largo."	AGN Tierras, vol. 2676, Exp 12
1571	"Me sembraba cada macegual seys brazas de sementera de ancho y ciento de largo."	AGN Tierras, vol. 60, Exp. 2, F44r-55v
1571	"Hazíale cada macegual seys brazas de simentera de ancho y ciento de largo."	AGN Tierras, vol. 60, Exp. 2, F75r-87v
1579	"…las diez brazas en cuadra que por esta Real Audiencia se les manda hazer para su comunidad…"	AGNP Protocolo de Tepeaca, paq. 41, exp. 63
1593	"Venden a Juan de Molina un pedazo de tierra de pan sembrar 100 brazas de ancho y 200 en largo en el pago de Coyotepetl"	AGNP Protocolo de Tepeaca, paq. 3, exp. III, f. 648
1587	Don Juan de Aquino sells a piece of land 70 by 40 *brazas* for the payment of San Nicolás Tamazula	AGNP Protocolo de Tepeaca, paq. 2, exp. I, f. 9r

FIGURE 72. EXTRACTS FROM TEPEACA HISTORICAL SOURCES SHOWING THE SIZE OF LAND ALLOTMENTS TO THE TERRAZGUERO SECTOR.

literally meaning "me in the shape of a point", that is, upright or erect. Sometimes the word *maytl* (arm) is put as a prefix, perhaps to emphasize that the measure involves raising the arm (see examples in Matías 1984). This type of vertical fathom was widely used in various parts of central Mexico, but recurrently is referred to under the corrupt terms of *neuitzan*, *neguitzantles* or *niquitzantlis* (Matías 1984: 13-22) (Figure 73). For Tepeaca, I have not found any reference that indicates its equivalent in terms of Spanish *varas*. However, its value equals the distance computed from the foot to the tip of the fingers with the arm raised above the head. It is not entirely clear whether the measurement is taken crossways, as specified in the *Libro de las Tasaciones* for Tultitlán (Gonzalez de Cossio 1952): 'measure of each braza from the left foot to the right hand with arm raised'. Since I have only found only one mention of this kind, for the moment we may assume that generally the measure was taken from the foot to the hand, both from the same side.

A text from northwest Puebla clearly states that the *nehuitzantli* measured 2.5 Spanish *varas* or around 2.09 meters.[vii] Obviously, this is only an approximation because the value had to vary according to the different individual body dimensions. Yet, this equivalence again corresponds well with the proportions of a person with an average stature of 1.67m and who could reach just above 2 m when raising his arm above his head (Leander 1967: 36). However, texts from other areas hint that the *nehuitzantli* might have had a higher value. One example is the famous description by Alva Ixtlilxochitl regarding the dimensions of the Palace of Nezahualcoyotl in the Tezcoco region of the Basin of Mexico:

The houses had in length and ran from east to west, four hundred and eleven and a half measures, which reduced to our measure make up one thousand two hundred and thirty four and a half varas, and in latitude which is north to south, three hundred and twenty six measures that make up nine hundred and seventy eight varas (Alva Ixtlilxochitl 1997:cap xxxvi, pp. 92-93) (my translation).

This reference has led Victor Castillo (1972: 212-213) to conclude that such measures are vertical *braza*. He assigned a value of 2.50m to it because each measure equals three *varas* de Burgos (3 x 0.8359). The problem is that Alva Ixtlilxochitl only equated three Spanish *varas* with one unknown measure, which is presumably of indigenous origin. This confusion is augmented by the fact that the Spaniards commonly named any measurement that involved the use of the arms as a *braza*. Examples of this type of confusion can be seen with the use of the terms *cemmitl*, *cemmatzotzopaztli* and *cemmatl* described by Alonso de Molina (2008 [1571]). Another possibility is that the cited measure, as Williams and Harvey (1997: 26-27) indicate, may refer to the *tlalquahuitl* or indigenous rod. This is an instrument used by the person in charge of measuring plots which, as I show below, could include different values (see below). With all probability, Ixtlilxochitl's unnamed measurements are *tlalquahuitl* and not *nehuitzantli*.

In the town of Xochimilco, located in the southern portion of the Basin of Mexico, the *nehuitzantli* was also commonly employed. Lockhart (1992: 145) found in a 1568 document that it was defined as a *braza* measured from the foot to the hand. He assumed that it made reference to a length of around 2.10 and 2.40m. Also, Matías (1984: 23-24) provides two measurements for the same locality, one from 1600 and a second from 1650. Its value was placed

---

[vii] The measure was obtained "…del pie a la mano y habiéndose medido la tal braza…con una vara de medir paños y sedas pareció que cada braza …tiene dos varas y media" (Archivo General de la Nación, Tierras, vol.73, Exp.5, f.171).

Date	Text	Source
1557	"...ciento y veynte brazas de largo y ciento de ancho, que se entienden de pie a mano..."	AGNP Protocolo de Tepeaca paq. 1, exp. 4, F 9r-11v
1557	"...ciento e veynte brazas de largo, que se entiende de pie a mano, e ciento en ancho..."	AGNP Protocolo de Tepeaca paq. 1, exp 4, F 11v-14r
1563	"...yn tlalli...Tecamachalco auic inic patlahuac napolneuitzantli ypan chicueneuitzantli ypan cematl ynin yneuitzan yn omoteneuh Francisco Quetzpan..."	AGN Vínculos, vol 60, exp. 3, f. 18v.
1580	"...una sementera de maíz de cuatrocientas brazas en cuadra del pie a la mano..."	Relación de Tepeaca
1585	A portion of land 80 by 84 brazas from the foot to the hand	Protocolo de Tepeaca, paq. 40, exp. 35
1597	"...700 brazas del pie a la mano de la medida que los naturales llaman niquizantlis..."	Protocolo de Tepeaca, paq. 40, exp . 2-R, ff. 719-720
1603	"...40 brazas del pie a la mano, que en lengua de los naturales se llama neguizantles, de largo, y de ancho 30 neguizantles."	Protocolo de Tepeaca, paq. 5, Exp. 3, ff. 15r-5v

FIGURE 73. TEXTS SHOWING THE USE OF THE NEHUITZANTLI IN THE TEPEACA AND TECAMACHALCO WITHIN THE STUDY REGION.

as 3 *varas* (2.51m) and 3 1/2 *varas* (2.93m) respectively. However, it is unlikely that the *nehuitzantli* of an indigenous person of 1.67m in stature could reach more than 2.50m. In order for this to be correct, the average height of the native population would have to be around 1.90m, for which there is no evidence. Even if someone occasionally attained 1.90m in height, it would have been so atypical that it is unlikely his *nehuitzantli* would have been taken as a reference measurement. Also, the 1650 equivalence of almost 3m is definitively suspicious. It may be that there is some kind of mistake made by the Spanish scribe, or that the Spanish vara employed had a different length than that of the *vara de Burgos* (e.g., *vara* de Burgos, *vara* de Aragón). As mentioned above, the value for the *vara de Burgos* is 0.8359 meters. The *vara de Toledo* has a very similar value to that of Burgos, and might actually be the same measure but with a different name, while the *vara* de Aragon was certainly shorter gauging 0.772m. However, given that the vara de Burgos was already established as the standard measure, I am inclined to think that such report was either (1) intentionally done in order to claim more land or to extract more tribute from the peasant community or (2) the Spanish scribes were confused and named these field measurements *nehuitzantli* instead of *tlalquahuitl*. Therefore it is more congruent to establish the value of the *nehuitzantli* at 2.09m for a person of 1.67m in stature. This is the value that I believe was employed in the Tepeaca area.

The indigenous rod or tlalquahuitl

The phrase 'indigenous rod' for measuring fields appears sparsely in the Tepeaca historical documents. It probably occurs as a direct translation of the Nahuatl word *tlalquahuitl*. As in other parts of central Mexico, it had various lengths, and for Tepeaca it had at least two values. In one document, it is specified that the rod was made up of two *braza* (Martínez 1984b: 81), which is the same equivalence for partitioning land in the town of Toluca according to data provided by Alonso de Zorita (1942). Another citation from the Protocolo de Tepeaca (page 41, file 131) it is stated that each indigenous rod had five rods for measuring cloth and silk [vara de Burgos].

Outside the Tepeaca region, the *tlalquahuitl* also has different values assigned to it. The *Libro de las Tasaciones* (González de Cossío 1952: 492-493), speaks of an allocation of land to the town of Tixtla, Tlaxcala. Here the Indians were '...obliged to work a maize field, and it was not declared of what quantity, and now the mentioned judge staked and measured it, which it has two hundred and forty two rods of three *braza* each rod in length, and one hundred and twenty eight wide...'. Thus, in this situation the value of the *tlalquahuitl* is around 5.01m (1.67m x 3 *braza*).

Occasionally, the *tlalquahuitl* is also equated with other measures such as the *nehuitzantli*. A 1604 Nahuatl text from the community of Cuahuixmatlac Atetecocho, Tlaxcala, translated by Luis Reyes (2001: 24-26), explains that Juan Citlalpopoca and Cristina Caxtilanxochitl declared that '... we have sold our land, our cultivating land to the principal lady María de Morales... which has in height three hundred and one rods for land [*tlalquahuitl*] or *nehuitzantli* which is the measure that is taken from the foot to the hand...' (my translation to English). The original Nahuatl text says 'on ce tlalquahuitl caxtolpohuali nehuitzantli' so it might actually be saying 'three hundred and one *tlalquahuitl* the length of a *nehuitzantli*'. That is, the rod has the length of a vertical *braza*. Similar to this, is the measure for fields described in a document from Mexico dated to 1592 in which it is stated that a *tlalquahuitl* equals '...a rod with the length from the foot to the hand according to the order in which the natives measure'.[viii] Perhaps, too, the land *braza* reported by Hans Prem (1978: 294) for Huexotzingo which equaled three rods for measuring cloth and silk,

---
[viii] Archivo General de la Nación, Tierras vol. 55, Exp. 2, fr. 26r.

and the previously mentioned indigenous measure of Alva Ixtlilxochitl, in fact, are values established for the *tlalquahuitl*.

What this indicates is that the *tlalquahuitl* was an instrument used for recording the perimeter of the parcels. For this reason, the 16th century AD Nahuatl-Spanish dictionary of Alonso de Molina (2008 [1571]) translates it as a 'measure for land with which it is measured ' or 'rod for measuring lands or estates'. Its length could be determined based on any of the standardized forms such as the *nehuitzantli*, *cemmitl* or *cenyollotli*, or by multiples of them. Consequently, its length was highly variable and could range from two to five horizontal fathoms (~3.34m to ~5.01m), to five Spanish *varas* (~4.17m), or have the length of a *nehuitzantli* (~ 2.09m).

The size of agricultural plot

The detachment of land for use by political institutions from the subsistence base of *macehualli* populations can be established by calculating the production capacity of the local Tepeaca agricultural systems and the amount of area worked by individual households. As mentioned above, nobles provided each tenant with five to nine parcels for rent, each 6 *braza* wide by 100 in length. One of these was destined for the payment of terrazgo rent, which totaled 600 square *braza*. If the value of the horizontal *braza* or *cemmatl* averaged 1.67m, then the farmed area to pay the rent totaled around 1673m^2.

On the other hand, area cultivated by vassals of the *Relación de Tepeaca* was 400 *nehuitzantli*, for 400 individuals. This translates into 160,000 squared *nehuitzantli* as the area cultivated. Dividing this area by 400 laborers indicates that each person cultivated the equivalent of 400 squared *nehuitzantli*. If the average length of the *nehuitzantli* was 2.09m^2, then 400 squared *nehuitzantli* would represent about 1747m^2.

Interestingly, the values expressed in *nehuitzantli* can be converted to *cemmatl* or vice versa using the relationship of the human body proportions. The *cemmatl* of an individual with a stature 1.67m represents 80% of his *nehuitzantli*. That is, for each 4 *nehuitzantli*, 5 *cemmatl* are obtained. Therefore, the tribute area for the vassals of the *Relación de Tepeaca* totaled 250,000 square *cemmatl*, which divided by the 400 vassals, totals 625 squared *cemmatl* for each one. The resulting data are revealing. The amount of land allocated as tribute payment is virtually the same for both cases. There is only a difference of 25 squared *cemmatl* (~70m^2) per individual. Furthermore, this residual may be due to conversion factors between the two types of measures. Clearly, either measure could be used depending on individual decisions or the needs of each field partition.

The area destined for the payment of tribute was relatively small. The *macehualli* only had to till about 0.17ha or 10%-15% of the area they were provided. This area seemed not to have surpassed 600 squared *braza* (*cemmatl*), and in some instances was even lower as it is exemplified in two documents from Tepeaca. In the Memoria of Don Diego Ceinos dated to 1571, several merchants had to cultivate one parcel 3 by 80 *braza* (*cemmatl*) (669m^2) as tribute payment, while in the Memoria of Diego Olarte, also dated to 1571, it is indicated that four bricklayers tilled an area of 2 x 100 *braza* (*cemmatl*) (558m^2) (Martínez 1984a: 447-448, 454-455).

The allocation of relatively small spaces is understandable considering the limited labor available within the household. Given prehispanic agricultural technology, a household with 5 to 7 people could only work a little more than one hectare of land annually. The main restraints were a shortage of labor within individual households, the simple agricultural technology, the lack of traction animals, and the poor transportation systems (Sanders and Webster 1988). The optimal strategy for the nobility was to impose a limited agricultural tribute obligation. This gave *macehualli* the best opportunity to meet their household consumption needs and thus ensure their survival. Placing excessive labor demands on households would have severely compromised their livelihoods and reduced the available labor for both agricultural and craft production.

Production capacity at the subsistence and institutional agriculture level

An estimate of the productive potential of ancient Tepeaca agriculture is crucial for analyzing the social relationship between the indigenous political and commoner sectors of the Tepeaca region. For this, I consider two basic premises. First, maize was the main staple in the indigenous diet. Other domesticates also formed an important proportion of the diet including amaranth (*Amaranthus*), beans (*Phaseolus*), squashes (*Cucurbita*), and chili (*Capsicum*) (Torres 1985). In particular, amaranth was a cereal regularly consumed in the prehispanic period and provided a similar nutritional value as maize. Both maize and amaranth are C4 plants that have higher yields than other cultigens. Today, some farmers in the Tepeaca region cultivate amaranth and consume it as a green vegetable and as bread named *alegría*. Unfortunately, we lack detail information on how much amaranth was used among indigenous populations; its use diminished considerably during the Colonial Period, probably because the Spaniards associated it with idolatry (Torres 1985: 71-74). Yet, given that maize and amaranth have a similar caloric contribution to the indigenous diet, I consider the former to be a useful proxy for estimating agricultural productivity and the impact of climate variability on food production systems.

Secondly, no matter what environmental conditions prevailed in the Tepeaca region 500 years ago, prehispanic agricultural systems were undoubtedly constrained by local climatic phenomenon. If Late Postclassic and Early Colonial environmental conditions were similar to those present today, then we can assume that most *macehualli* populations practiced rainfall agriculture (RA) as the primary form of cultivation.

Institutional agriculture production

Nobles controlled large territories and their tributary populations. Their domains comprised a single continuous geo-political unit with loose frontiers and intertwined territories (Martínez 1984b: 27-44, 53-55). The boundary areas between populations within a single *altepetl* were blurry and the houses of one town were commonly entangled with those of the other (Cook 1996: 112). However, territorial limits between each political entity were firmly established. The 16th century AD boundaries of the Tepeaca *altepetl* were portrayed in the Cuauhtinchan Map 4 (Yoneda 1994). This indicates that during the Colonial period elite obtained agricultural tribute only from the *macehualli* inside their respective polity. In contrast, before the Spanish arrival, military expansion allowed the Tepeaca political institutions to extract resources from other conquered populations. We know that during the prehispanic period tribute was collected from several areas within the Tepeaca territorial domain, which extended as far as Iztacamaxtitlan and Xalacingo to the north, Matlatlan and Acultzingo to the east, and Huehuetlan, Zapotitlan and Tepexi to the south (Martínez 1984b: 54).

Tepeaca institutional agriculture had the ability to mobilize a large quantity of food resources. In the Colonial period, the control of large landholdings allowed nobles to obtain and mobilize important amounts of tribute produced by *macehualli/terrazguero* households. To exemplify how institutional agriculture operated at the local level, let us examine the case of Doña Francisca de la Cruz, one of the Tepeaca's five tlahtoani during the early years of Spanish rule. Doña Francisca owned lands in at least 53 different barrios or estancias for which we have detail information on 25 specific landholdings (Martínez 1984b: Cuadro 2). In her 1581 memoirs, she declared having 1610 *macehualli/terrazguero* distributed in 25 landholdings within the Tepeaca *altepetl* (see Perkins 2007: Table 1). Most of tributaries were located in settlements near Acatzingo and a few bordered the adjacent Tecamachalco, Tecali and Cuauhtinchan *altepemeh*. Agricultural fields were situated in barrios, estancias, or parajes whose names are still in use today. These fields are located by barrio name in Figure 74.

Each *macehualli/terrazguero* household attached to Doña Francisca was assigned between five and nine agricultural fields of 600 squared *braza* (0.17ha). Thus, the land she

Field	prehispanic Settlement	Modern Location	Total HH	Total number of fields (5-9 per HH)	Total Landholdings (ha)	Area destined for tribute (ha)
1	San Vicente Tlacaltech	Near Cuauhtinchan-Tecalco	100	500-900	85-153	17.00
2	Tlayalac	Barrio in Acatzingo	200	1000-1800	170-306	34.00
3	San Joachin Zayacatlohtlan	Barrio in Acatzingo	60	300-540	51-92	10.20
4	S. M. Purificación Chichico	Barrio in Acatzingo	40	200-360	34-61	6.80
5	San Mauricio Coyoal	Barrio in Acatzingo	60	300-540	51-92	10.20
6	San Felipe Quauazala	Barrio in Acatzingo	20	100-180	17-31	3.40
7	Chiquiuican San Bartolomé	Barrio in Acatzingo	40	200-360	34-61	6.80
8	San Antonio Tetopizco	Barrio in Acatzingo	20	100-180	17-31	3.40
9	Auatla	In Acatzingo	80	400-720	68-122	13.60
10	San Pedro y San Miguel	Barrio in Acatzingo	40	200-360	34-61	6.80
11	Santa María Nativitas Tetela	Barrio in Acatzingo	40	200-360	34-61	6.80
12	San Antonio Mecapala	Barrio in Acatzingo	20	100-180	17-31	3.40
13	San Juan Zotolocan	Barrio in Acatzingo	20	100-180	17-31	3.40
14	Acaxic	Barrio in Acatzingo	15	75-135	13-23	2.55
15	Ayahuaculco	Barrio in Acatzingo	20	100-180	17-31	3.40
16	Santo Tomás Chicanyocan	Near Nopalucan	80	400-720	68-122	13.60
17	Santa Catalina Yacapiztlan	Near Tecalco	40	200-360	34-61	6.80
18	San Salvador de Uzcalotla	In Huixcolotla	40	200-360	34-61	6.80
19	Atlamaxac San Lorenzo	Barrio in Acatzingo	20	100-180	17-31	3.40
20	Capola Santa Inés	Near Tecalco	60	300-540	51-92	10.20
21	Santa María Ocoyocan	Barrio in Acatzingo	50	250-450	43-77	8.50
22	Tzoquitzinco Santa Ana	Barrio in Acatzingo	35	175-315	30-54	5.95
23	Tianquitenpan	Barrio in Acatzingo	400	2000-3600	340-612	68.00
24	Ayapanco	Estancia in Acatzingo	100	500-900	85-153	17.00
25	Jalticpac	Barrio in Acatzingo	10	50-90	9-15	1.70
Total			1610	8050-14,490	1369-2463	273.70

FIGURE 74. LANDHOLDINGS OF DOÑA FRANCISCA DE LA CRUZ.

controlled was around 1369 to 2463ha. Given that the payment of the terrazgo proceeded from one field per household, then the effective area from which Doña Francisca extracted agricultural tribute was 273.70ha (1610 x 0.17 per household).

It is difficult to determine the average agricultural tribute volume given annually to Doña Francisca de la Cruz. Historical data lacks information regarding the amount of maize given as tribute. However, it is possible to generate a raw estimate of average maize yields for each of her landholdings. This can be made if we assume that the majority of households practiced rainfall agriculture and crops were affected by variable environmental conditions.

As mentioned in Chapter 6, average maize production under rainfall conditions for the study region averages 1430kg/ha. However, there are important differences in maize yields within each environmental zone and Class type lands (see Figures 56 and 57). Taking this into consideration, I have assigned an average yield value for each of Doña Francisca's 25 landholdings according to their geographic location within the Tepeaca region (Figure 75). Maize yields employed for this estimation were obtained from modern fields that were fertilized with animal manure. For this reason, it can be argued that current maize yields are higher than those during ancient times. However, as I pointed out in Chapter 2, prehispanic agriculturalists employed several natural fertilization methods directed at maintaining adequate soil fertility. Such methods included the application of night soil, forest humus, and house refuse to the fields, as well as plant intercropping techniques. Given that lack of concrete data, I consider that under rainfall conditions modern and Late Postclassic maize yields were similar.

Doña Francisca's fields would have had contrastingly different yields ranging from as much as 2206kg/ha in Class 2 lands in the Nopalucan area, to as little as 346kg/ha in marginal Class 5 lands in the southern town of Uzcalotla (modern Huixcolotla). In the middle, she would have had moderately yielding Class 3 and 4 lands in the Tecali-Cuauhtinchan border (1056kg/ha) and Class 5 lands in Acatzingo (1075kg/ha). Considering that each *terrazguero* household worked an area of 0.17ha as the tribute payment, then Doña Francisca received annually around 304 metric tons of maize from her 25 landholdings (Figure 76). Furthermore, assuming arbitrarily that between years the average maize productivity experienced an increase or decrement of 50% due to variable climatic conditions, then maize acquired annually by way of tribute would have ranged between 162 and 487 metric tons (Figure 77).

Producing and collecting large quantities of maize was important because the *teccalli* financed other complex political institutions during the Late Postclassic period. In addition, noble lords (*teuhctli*) supported a series of lesser rank nobles (*pilli*). Nobles were certainly fewer in number in relation to the *macehualli* population, although probably not as low as has been suggested by John Chance (2000:

Landholding	Location	Average maize yields (kg/ha)
1, 17, 20	Near the Cuauhtinchan-Tecali border	1056*
2-15, 19-25	In Acatzingo and its vicinity	1075
16	Near the Nopalucan border	2206
18	In Huixcolotla, near the Tecamachalco border	346
* Averaging the Cuauhtinchan and Tecali historical average yields		

FIGURE 75. ASSIGNED AVERAGE MAIZE YIELDS IN DOÑA FRANCISCA DE LA CRUZ 25 LANDHOLDINGS.

448). We know that Doña Francisca de la Cruz had at least 48 dependents, 46 of which were males. This marked difference in numbers between genders suggests that the individuals recorded were the heads of noble households. Considering that each household had at least 2-3 members elevates the number of people that she supported. If agricultural production under an average agricultural year (between 152 and 456 metric tons) were divided equally among the 48 noble households in Doña Francisca de la Cruz *teccalli*, then each would have received between 3.16 and 9.50 (6.33 metric tons on average) of maize annually, an amount well above their subsistence needs.

Unless noble households were extraordinary large, each one was able to amass a substantial amount of maize surplus. The quantity that Francisca de la Cruz obtained annually by way of agricultural tribute was probably much higher because we only have partial information on her landholdings. More resources were probably obtained from the *macehualli* sectors that possessed patrimonial lands, and from particular types of properties that the political institutions controlled. In the Tepeaca-Cuauhtinchan region, such properties include the patrimonial lands (*neixcauilaxca*), the ruler's lands (*tlatocacuemitl*), the nobility lands (*pilcuemitl* or *couacuemitl*), and the land of the noblewoman (*teuccihuacuemitl*) (Reyes 1988a: 122). In the Late Postclassic, the agricultural tribute was extremely large because tribute was extracted from both *macehualli* using patrimonial lands and from conquered populations. Unfortunately, we lack the information needed to calculate this additional resource extraction.

How efficiently could the Tepeaca nobles cope with climatic variability and severe droughts? To answer this question I examine a severe drought and crop failure scenario using the 2009 canícula as a proxy to calculate effects (see Chapter 6). This exercise is useful for comparing maize production for an average production year with a bad one. For this, we must assume that the effects of severe droughts on crop stands were similar in past times as they are today.

Field	Settlement (barrio, estancia, paraje)	Number of terrazguero households	Total area destined for tribute (ha)	Average maize productivity (kg/ha)	Maize production acquired as tribute (metric tons)
1	San Vicente Tlacaltech	100	17.00	1056	17.95
2	Tlayalac	200	34.00	1075	36.55
3	San Joachin Zayacatlohtlan	60	10.20	1075	10.97
4	Sta. María Purificación Chichico	40	6.80	1075	7.31
5	San Mauricio Coyoal	60	10.20	1075	10.97
6	San Felipe Quauazala	20	3.40	1075	3.66
7	Chiquiuican San Bartolomé	40	6.80	1075	7.31
8	San Antonio Tetopizco	20	3.40	1075	3.66
9	Auatla	80	13.60	1075	14.62
10	San Pedro y San Miguel	40	6.80	1075	7.31
11	Santa María Nativitas Tetela	40	6.80	1075	7.31
12	San Antonio Mecapala	20	3.40	1075	3.66
13	San Juan Zotolocan	20	3.40	1075	3.66
14	Acaxic	15	2.55	1075	2.74
15	Ayahuaculco	20	3.40	1075	3.66
16	Santo Tomás Chicanyocan	80	13.60	2206	30.00
17	Santa Catalina Yacapiztlan	40	6.80	1056	7.18
18	San Salvador de Uzcalotla	40	6.80	346	2.35
19	Atlamaxac San Lorenzo	20	3.40	1075	3.66
20	Capola Santa Inés	60	10.20	1056	10.77
21	Santa María Ocoyocan	50	8.50	1075	9.14
22	Tzoquitzinco Santa Ana	35	5.95	1075	6.40
23	Tianquitenpan	400	68.00	1075	73.10
24	Ayapanco	100	17.00	1075	18.28
25	Jalticpac	10	1.70	1075	1.83
Total		1610	273.70	1089 (average)	304.01

FIGURE 76. ESTIMATED MAIZE PRODUCTION UNDER AN AVERAGE YEAR IN DOÑA FRANCISCA DE LA CRUZ'S LANDHOLDINGS.

Field	Average maize yields (kg/ha)	50% decrease	50% increase	Total area destined for tribute (ha)	Range of maize production acquired as tribute (metric tons)
1, 17, 20	1056	528	1584	34.00	17.95-53.86
2-15, 19-25	1075	538	1613	219.30	117.87-353.62
16	2206	1103	3309	13.60	15-45
18	346	173	519	6.80	1.18-3.53
Total				273.70	152.00-456.01

FIGURE 77. RANGE OF MAIZE PRODUCTION ASSUMING A 50% INCREASE OR DECREMENT IN AVERAGE MAIZE YIELDS IN DOÑA FRANCISCA DE LA CRUZ LANDHOLDINGS

I have plotted Francisca de la Cruz's 25 known fields over the 2009 regional maize production map of the Tepeaca region (Figure 78). The location of her fields is only an approximation, but it provides a way to evaluate field dispersion strategies in relation to overall productivity. For each landholding, I assigned the correspondent maize yield relative to its location on the maize production map generated with the 2009 ethnographic data. I then multiplied the productivity value of each plot by the number of farmers in each site (Figure 79).

The results indicate that under a major drought similar to the 2009 dry canícula event Doña Francisca de la Cruz would have been able to accumulate around 45 metric tons of maize. This represents only 14.80% of an average (304.01 metric tons) year's production. With such production, each of the 48 noble household dependents of de la Cruz's would have received around 937.50kg of maize. This amount would have been enough to survive, but nothing more. In consequence, a severe climatic disruption represented a considerable blow to the accumulation of agricultural

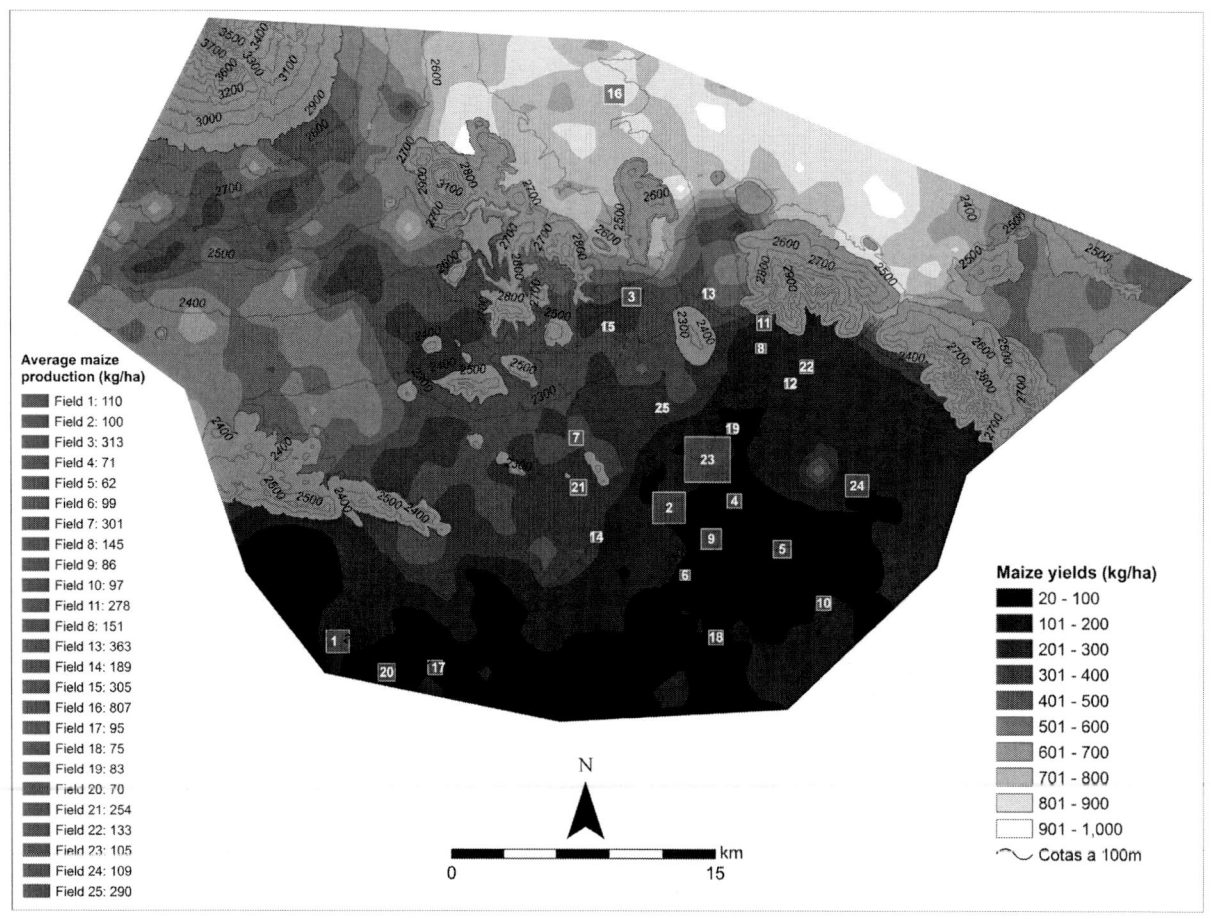

FIGURE 78. MAP OF THE 2009 MAIZE PRODUCTION DISTRIBUTION IN THE TEPEACA REGION. OVER IT, ARE POSITIONED THE APPROXIMATE LOCATION OF DOÑA FRANCISCA DE LA CRUZ'S LANDHOLDINGS.

surplus in comparison to a regular year. However, since nobles could store large amounts of maize from year to year, one or two years of crop failure probably represented a minor problem easily buffered.

Subsistence agriculture production

Tepeaca subsistence agriculture had important limitations. *Macehualli/terrazguero* cultivated small plots (0.68 to 1.36ha) using predominantly rainfall agriculture in an area characterized by variable climatic conditions and unstable crop yields. Yet 16th century AD Tepeaca nobles might have considered that such conditions were sufficient to sustain *macehualli* populations and did not interfere with tributary payments of staple and wealth products. Both nobles and commoners must have acknowledged the intrinsic unpredictability of agricultural production. From the information presented it is logical that households could not depend only on a single year's production. A certain level of surplus had to be produced in order to compensate for food shortages that could arise in other years. Although subsistence agriculture was destined mainly for the reproduction and survival of household members, *terrazgueros* probably sought to produce some agricultural surplus to secure their survival in poor years. Doing so was also critical for the political institutions in order for the tributary system to work. Without the commoner population, the Tepeaca elite could not be supported in the manner that they desired.

It is relevant then to examine the capacity and degree of agricultural surplus at the household level. To do so, I estimate the capacity of *terrazgueros* to produce maize according to the available land and the local environmental conditions. I then compare the food requirements with estimated maize production of households from four locations within the study region. Internal differences in food production may have existed within the community and most likely *macehualli* households varied in size and had unequal access to land resources. An unequal distribution of land may have existed in Tepeaca during the Late Postclassic when settlements had high population densities (Cook 1996: 113; Anderson 2009: 187, Figure 5-44). Some households were given a minimum of five plots (0.68ha) and others up to nine (1.36ha). Given the lack of data, I will assume that on average households required one metric ton of maize for their annual consumption needs (Sanders 1976a).

Maize production within macehualli/terrazguero households

Terrazguero households cultivating in different environmental conditions probably had contrasting maize

Field	Settlement (barrio, estancia, paraje)	Total of *terrazguero* households	Total area destined for tribute (ha)	Maize productivity during a *canícula* event (kg/ha)	Maize production acquired as tribute (metric tons)
1	San Vicente Tlacaltech	100	17.00	110	1.87
2	Tlayalac	200	34.00	100	3.40
3	San Joachin Zayacatlohtlan	60	10.20	313	3.19
4	Sta. Ma. Purificación Chichico	40	6.80	71	0.48
5	San Mauricio Coyoal	60	10.20	62	0.63
6	San Felipe Quauazala	20	3.40	99	0.34
7	Chiquiuican San Bartolomé	40	6.80	301	2.05
8	San Antonio Tetopizco	20	3.40	145	0.49
9	Auatla	80	13.60	86	1.17
10	San Pedro y San Miguel	40	6.80	97	0.66
11	Santa María Nativitas Tetela	40	6.80	278	1.89
12	San Antonio Mecapala	20	3.40	151	0.51
13	San Juan Zotolocan	20	3.40	363	1.23
14	Acaxic	15	2.55	189	0.48
15	Ayahuaculco	20	3.40	305	1.04
16	Santo Tomás Chicanyocan	80	13.60	807	10.98
17	Santa Catalina Yacapiztlan	40	6.80	95	0.65
18	San Salvador de Uzcalotla	40	6.80	75	0.51
19	Atlamaxac San Lorenzo	20	3.40	83	0.28
20	Capola Santa Inés	60	10.20	70	0.71
21	Santa María Ocoyocan	50	8.50	254	2.16
22	Tzoquitzinco Santa Ana	35	5.95	133	0.79
23	Tianquitenpan	400	68.00	105	7.14
24	Ayapanco	100	17.00	109	1.85
25	Jalticpac	10	1.70	290	0.49
Total		1610	273.70	188 (average)	45.00

FIGURE 79. MAIZE PRODUCTION ESTIMATE FOR DOÑA FRANCISCA DE LA CRUZ FIELDS UNDER AN ASSUMED CANÍCULA DROUGHT EVENT.

yields. To examine these differences, I compare four potential scenarios of maize production from households located in four landholdings of Doña Francisca de la Cruz (Fields 1, 9, 16 and 18; see Figure 78). I have chosen these four locations because they are located in contrasting environmental settings with different levels of maize productivity. One of Doña Francisca's lands was located in San Vicente Tlacaltech (Field 1) near the Cuauhtinchan-Tecalco border in the southwestern corner of the Tepeaca Valley. Agricultural fields were set on Class 3 lands which today have a relatively good average maize productivity of 1056kg/ha (averaging the Cuauhtinchan and Tecali historical maize yields).

The second landholding was located at Auatla (Field 9) in the central portion of the Tepeaca valley. Land in this area was situated in a Class 5 zone with poor environmental conditions for agriculture and an average maize yield of 1075kg/ha.

The third land, Santo Tomás Chicanyocan (Field 16), was located near the boundaries with Nopalucan in the Llanos de San Juan Region. Households would have had access to Class 2 lands with a good potential for agriculture and an average maize productivity of 2206kg/ha.

The fourth land was located in San Salvador Uzcalotla (Huixcolotla) (Field 18) in the southeast portion of the Tepeaca valley. Households here would have had access to Class 5 lands located in marginal agricultural zones. Soils in this area are the poorest within the study region. In addition, rainfall is scarce and highly erratic between years. This is reflected in low average maize yields of only 346kg/ha.

The range of maize produced by a *macehualli/terrazguero* would have contrasted from household to household. Households in Santo Tomás Chicanyocan had a higher potential for surplus accumulation. As shown in Chapter

6, the Llanos de San Juan have much better environmental conditions for agriculture, including deeper soils with more humidity retention, compared to other areas in the Tepeaca valley. Regardless of intrinsic climate effects, the region's optimal growing conditions allow higher average maize yields. The range of production for *macehualli/ terrazguero* households in Chicanyocan would have fluctuated between 1500 and 3000kg/ha.

In an average year, households from Tlacaltech, Auatla and Chicanyocan would have been able to produce enough maize for their requirements. Maize production in Tlacaltech (718-1436kg) and Chicanyocan (1500-3000kg) probably was well above the annual requirements of households giving them the possibility to generate surplus. This allowed *terrazgueros* to store a portion of the year's production for future use, and probably promoted the exchange of maize for other non-perishable products.

Households in the Uzcalotla (Huixcolotla) area were at a constant risk of production shortfalls and food shortages. The main restraint was the low agricultural productivity of the region (346kg/ha) due to poor soils, low precipitation levels and a high evo-transpiration rate. Under such conditions, household agricultural production would have constantly fallen below yearly needs, an effect that became more pronounced in years of harsh climatic disruptions or under plague affectations. Uzcalotla probably exemplifies much of the conditions of the southern and southeastern portion of the Tepeaca valley were households cultivated in marginal agricultural areas. Locations such as Caltenco, Tecamachalco, and Tochtepec probably faced similar production difficulties. These zones today report extremely low average maize productivity below 350kg/ha. In such circumstances, inputting more labor to agricultural tasks was probably not a key factor determining the level of food production and economic well-being. As much as households could provide more labor, poor environmental conditions and the lack artificial water supply to lands put a limit on the productive potential on local agricultural systems. As a result, food deficits were probably a regular component of adaption in these areas requiring households to acquire food on a regular basis via the exploitation of other resources.

In most cases, it is probable that *macehualli/terrazguero* were capable of producing some maize surplus over the long run. Occasional total crop failures (zero productivity) may have forced households to acquire food via other sources or through exchange mechanisms. However, local food shortages could have been dealt with if households stored and kept a good reservoir of maize. Modern farmers from the study region usually store maize for one or two years after harvest, depending on conditions of preservation. Keeping it for a longer period is risky because some amount is lost due to rotting or plagues (Smyth 1989). Therefore, at some point or another, *terrazgueros* had to exchange maize or other products for non-perishable items during years of good production. These items were extremely important because they could be used in times of need. In addition, a portion of maize could move through reciprocal exchange between community members or used to acquire wealth goods for tributary purposes.

In areas with low agricultural productivity, and land limitations, restrained agricultural production of *macehualli/terrazguero* households and would have lowered the level of well-being as they became even larger. Households in these circumstances may have opted to engage in other economic activities to compensate for food deficits. Possible adaptations in marginal agricultural areas may have been early household fissioning. As soon as some members reached adulthood, they could form their own household and petition for lands. A second alternative would be to intensify production by inputting more labor per unit of space. This could involve putting more care into plants, especially in fields located near the homestead such as calmilli houselots (Killion 1990; Palerm 1955). Techniques involved in intensifying the care of plants include pruning, mulching, plant transplanting (Wilken 1987) and the manual suppression of ear prolificacy in maize. Yet, it is difficult to quantify how much productivity could be augmented in this way. A third option could have been to engage in alternative economic activities. As is shown below, domestic craft production was probably one of the most efficient ways to compensate for food deficits. Crafting activities were also a main activity of *terrazguero* households and items produced formed an important part of the tributary payment to local elite.

In sum, subsistence agriculture in ancient Tepeaca was characterized by relatively small spaces for cultivation and variable climatic conditions. These were critical factors that probably restrained *terrazguero* households for increasing production. However, there were probably substantial regional differences in the amount of maize produced annually. Areas with better soils and more stable climate conditions such as the Llanos de San Juan and the Amozoc-Acajete area probably had better yields and a greater ability to generate a food surplus. The southeastern area of the Tepeaca Valley was especially susceptible to annual food shortages. At the regional scale, subsistence agriculture was geared towards sustaining household members. Providing a stable subsistence base allowed the household to engage in other economic activities to produce items that could be exchanged in times of need for additional food resources. However, it is not surprising to see households in marginal conditions engage in a greater number of economic activities considering that agricultural productivity was poor and risky.

Buffering strategies against climatic variability

Field dispersion

The subsistence and institutional agriculture in Tepeaca shared basic principles. The most salient was the use of a field dispersion strategy in order to minimize crop failure. They differ mainly in the scale at which field dispersion was employed. With institutional agriculture, landholdings were separated by large distances that frequently surpassed 20km. This allowed nobles to take advantage of several environmental zones, making optimal use of the region's soils and precipitation regimes. It also lessened the risk of total crop failure due to seasonal shifts in rainfall. In Colonial times, nobles obtained food resources at different time intervals from the core Tepeaca territory (around 1300km^2) because of differential maize harvest times (see Chapter 6). Both conditions provided nobles with a steady supply of agricultural products throughout the year and gave them the capacity to amass substantial levels of surpluses during average years.

At the subsistence agriculture level, field dispersion was probably an effective strategy to buffer the risk of total crop failure. Indigenous populations favored an infield-outfield system. It had the advantage that intensive labor could be applied to infield near houselots (*calmilli*), which could be cultivated year-round due to their closeness to the homestead. *Calmilli* are more productive in than outfields. A large diversity of plants is cultivated in *calmilli* that complement in important ways the diet such as tejocote (*Crataegus mexicana*), capulín (*Prunus capuli*), tuna (*Opuntia* spp.), pitaya and pitahaya (dragonfruit cactus —*Hylocereus* spp.). Vegetables and spices are also easily grown near houses like squashes (*Cucurbita pepo, C. radicans*), tomatoes (*Physalis philadelphica*) and chile (*Capsicum annuum*).

Heath Anderson (2009: 277-281) thinks that the dispersed settlement pattern found in the Tepeaca region during the Late Postclassic period was in part the result of an infield-outfield cultivation strategy that included maguey cultivation as an important nutrient source. Because rainfed cultivation was the dominant form of agriculture in the region, maguey cultivation was probably an important nutritional component in infield and outfield cultivation. Maguey has the advantage of growing virtually anywhere. The plant will thrive in the most marginal landscapes and was considered a famine food by indigenous populations. Maguey has also many uses. In steep sloping terrains, rows of maguey (*metepantli*) are used as walls for soil retention and to prevent erosion (Patrick 1977; Wilken 1987). Maguey is also esteemed for its sap, which can be consumed directly as aguamiel or fermented into a mildly alcoholic beverage called pulque. The leaves of the maguey can also be roasted and consumed as a cooked vegetable. Its sap is high in calories and can serve as a complementary nutrient source for households. Susan Evans (1990) has shown that maguey cultivation was a critical component of household agriculture of dry marginal lands in Cihuatecpan, in the valley of Teotihuacan. Sap and vegetables crops would have produced more than 2,000,000 kcal/ha annually for households in Cihuatecpan (Evans 1990: 126). These calories would represent an important addition to families cultivating lands in marginal conditions that have a low maize productivity.

Comparable agricultural conditions to those reported by Evans (2001: 94, Table 2) in the Teotihuacan valley are found in the southeastern portion of the Tepeaca valley. Today, maguey is cultivated widely in towns with marginal agricultural conditions like Tecamachalco, Tochtepec, Caltenco, and Hueyotlipan. The main uses for maguey are for pulque production and the leaves for *barbacoa* or pit-roasting of animal meat. This would suggest that maguey sap consumption may have been an important activity for households located in these areas where maize production is low and risky. However, I believe that sap production was probably only one aspect of food procurement strategies in marginal or risky areas. Households probably diversified food acquisition exploiting a range of drought resistant resources. In El Palmillo, Oaxaca, marginal conditions for maize cultivation were no impediment for households which exploited a wide spectrum of semi-domesticated plants to support themselves (Feinman *et al.* 2007). In the Tepeaca region, there are various plants that thrive in semi-arid environments and can be exploited in different times of the year. These include huaje (*Prosopis laevigata*), izote (*Yucca periculosa*), biznaga (*Mammillaria* sp.), camote (*Ipomoea batatas*), and several species of *nopal* (*Opuntia* spp.). In addition, the southern arid areas of the Tepeaca valley are well known for good hunting of various species including rabbit, hare, reptiles, maguey and tree larvae, and until recently deer. Diversification of diet would have lessened the household's dependency on maize or other staple grains.

Outfields, as the name implies, are located away from the homestead and are usually worked less intensively due to the higher energetic costs to cultivate them. In general, farmers practicing traditional agriculture prefer to cultivate fields within 1-2km of their residence and rarely will cultivate fields beyond 4-5km from home (Chisholm 1970: 131). Beyond four kilometers, it becomes energetically inefficient to tend their fields due to associated travel time. In prehispanic times, *macehualli* households probably also preferred to cultivate as close to the homestead as possible in an attempt to minimize labor expenditure. During the Colonial period, farmers could have traveled larger distances because some traction animals were available. However, we know that horses and other animals used to carry cargo were largely restricted to nobles and merchants (Lockhart 1992) and not easily within the economic reach of commoners. Thus, the effective area of field dispersion was relatively low.

If *macehualli/terrazguero* cultivated fields within four-kilometer radius from their homes, then this would have allowed them to disperse their fields over around 50.27km^2. Such an extension could have helped to lessen typical seasonal fluctuations in localized rains, frosts and pest infestations. By diversifying risk both in terms of time (sowing at different times) and space (making use of several environmental zones), households could have averaged yields and avoided total crop failure. While this strategy can ameliorate many of the negative effects of localized climatic disruptions, it will also generate or create differential production levels between households.

The problem, of course, is that for households large scale extended droughts could have easily affected areas larger than 50km². In these events, a strategy of local field dispersion is not enough to prevent crop failure. Major climatic disruptions such as the dry canícula involve entire regions and ultimately produce agricultural food shortages at the regional level. Under such circumstances, food scarcity would have been felt homogenously across indigenous *macehualli/terrazguero* households and throughout the region.

Agricultural intensification

The most effective buffering strategy against climatic variability is to use Artificial Water Agriculture (AWA). Such systems employ irrigation techniques that are often linked to some form of institutional agriculture or elite involvement. However, agricultural intensification strategies have been well documented ethnographically among smallholders (Netting 1993; Stone 1996). In Tepeaca, lessening the negative effects of climatic variability could have been carried out by employing some form of irrigation. Irrigation technology was developed early in central Puebla. Formative archaeological features such as canals and drained fields have been recorded in Amalucan (Fowler 1987) and Cholula (Mountjoy and Peterson 1973), less than 30km west of the study region. Unfortunately the archaeological evidence for such systems is limited for the Postclassic period in Tepeaca (Sheehy, *et al.* 1995; Sheehy, *et al.* 1997). Poor preservation of prehispanic irrigation features may be part of the problem. However, local topographic conditions are not conducive for the development of irrigation and other AWA practices because of the region's highly porous sub-soils that prevent water accumulation on the surface (Medina 2001). In Early Colonial times, and up to the mid-19th century AD, most settlements captured and stored rainwater in small dams known locally as jagueyes:

> This city, the second seat which it presently has, is flat land. And, regarding its weather is of such quality that, at any time of year, the sun is the summer and the shade is the winter, though, simply, we call it cold land. And so it seems in all cases, especially because there is great difference between the trees and fruits of the cold and hot lands. And it is dry land because it is founded on limestone that, in some parts, the lime sprouts, and, in other parts, [it lies] at half a vara for measuring [in depth], and more or less. And this [condition] is known for the seat and trace that the city occupies, which has no river that passes by it, nor any spring of any quality. And so, in time of infidelity and after the Christians arrived, water was supplied from the rains, which they collected during the rainy season in earth rafts (called jagüeyes) with which they supplied [water] (Molina 1985 [1580]: 226) (my translation).

Springs were present in a few privileged areas. In Acatzingo, springs were used for both human consumption and small scale irrigation in haciendas and *rancherías* during the 18th and 19th centuries (Calvo 1973; Garavaglia and Grosso 1990). Both the commoner and the political sectors used spring water. However, the amount of water extracted was relatively small and could be used to irrigate only a very small fraction of the agricultural fields.

Neither today, nor historically, has the Tepeaca region had any perennial water streams. Several intermittent streams form in barrancas during the rainy season (See Figure 2). One of them, the Barranca del Aguila, carries a substantial flow of water that could have been used for irrigation. This barranca is formed by an interconnected system of faults that originate in the piedmont areas of the La Malinche volcano. Its course crosscuts the western portion of the Tepeaca valley and runs from north to south. Over most of its trajectory, except between Amozoc and Tepeaca, the barranca has a depth of more than 20m, making it difficult to exploit the water for irrigation. In the area near Oxtotipan (Oztoticpac) the barranca becomes a gorge reaching 100m in depth. In this zone, there are springs that feed the system all year long (Yoneda 2005: 247). However, as the depth of the barranca decreases near the town of Xochiltenango, water flow increases due to seepage in the porous travertine bedrock and water can be channeled effectively into the fields and used for irrigation (Medina 2001).

It is possible that the Barranca del Aguila system was exploited since the Late Formative period and may have represented a substantial intensification strategy for towns settled in the south strip of the study area. In the northern portion of the study area, the capacity of the barranca for intensification of production was probably restrained by two factors. First, the system depended entirely on rainfall patterns to capture water. Any fluctuation in precipitation in the region directly affected water availability downstream. Secondly, given that water flow is minimal during winter, irrigation would have been restricted to the summer period when it the barranca has its greatest discharge.

Because it is difficult to establish the extension of agricultural zones that could have been irrigated in prehispanic times, no attempt is made here to estimate its effect on maize production capacity. In any case, it is likely that any amount of irrigation in the region was probably controlled and managed, directly or indirectly, by the Tepeaca elite. This certainly would have given them the ability to generate larger agricultural surplus to finance stratification.

The marketplace

One particular strategy developed by the Mesoamericans, which I believe was highly efficient in the Late Postclassic and Early Colonial periods, was the local marketplace. During both the prehispanic and Colonial periods, the marketplace was an important locale where *macehualli*

	Diego Ceinos	Gabriel de los Ángeles	Melchor Rodríguez	Diego de Olarte	Francisca de la Cruz
Basket makers		4			2
Blacksmiths					3
Bricklayers				4	20
Carpenters					12
Feather workers					5
Hunters					7
Jeweler maker					2
Merchants	4	4		169	20
Painters	1			1	20
Petate makers		3	2	2	14
Potters		7	8		
Producers of tobacco			5	1	15
Sandal makers	4		5	4	15
Silversmith					5
Weavers					5
Total	9	18	20	181	145

FIGURE 80. NON-AGRICULTURAL TRIBUTARIES OF FIVE TLAHTOQUE FROM ACATZINGO AND TEPEACA. DATA OBTAINED FROM MARTÍNEZ (1984b: CUADRO 17).

populations obtained many necessary resources for their sustenance. They found that having to work their small fields in many cases required them to develop other alternative economic strategies for their support. If cultivation provided enough food for their annual subsistence needs, then small parcel cultivation enabled them to also pursue additional economic activities. In particular, multi-crafting and intermittent craft production were important income sources for increasing household well-being (Hirth 2009). It was also an efficient way to lessen the risks of food production shortfalls. For example, many of the Tepeaca's tlahtoque had a large number of tributaries engaged in a wide variety of alternative economic activities. The *terrazguero* tenants of Doña Francisca de la Cruz engaged in at least 16 different professions like bricklayers, carpenters, hunters, bead makers, basket makers, blacksmiths, potters, textile makers, merchants, perfume makers, *petate* makers, painters, silversmiths, feather workers, and shoemakers, some of which are listed in Figure 80. Yet, all of these tributaries had agriculture as a main sustenance activity. In addition, *terrazguero* women also engaged in cotton spinning and textile making as part of their tribute obligation. Surely, they also manufactured textiles for themselves and possibly for exchange.

Alternative subsistence activities besides agriculture were also important for mobilizing food resources. The *Relación de Tepeaca* indicates that crafting activities were extremely important in order to cover the expenses and needs of daily life:

And the Indians, as soon as they reach their maturity, have the same dealings and businesses, and are bakers, and all have and perform mechanical occupations, and, in their markets, they come with many different merchandize and foods, and, others, with their tents with things from Castilla and the [native] land, and they sow and harvest their maize fields and other legumes, which they buy and sell. And, the women of the natives, spin their cotton, and weave it and are benefited for their shirts and dresses, and many benefit and sell them, in canvas or as finished shirts (Molina 1985 [1580]: 257) (my translation).

In some cases, the need to produce items for exchange appears to have been a necessity. Some goods that the nobles demanded, such as cacao beans, had to be brought long distance to the Tepeaca marketplace. By producing alternate goods and not being entirely dependent on agriculture, *macehualli* could meet their tribute obligations and avoid harsh reprisals, such as those narrated by 16th century AD chronicler (Fernández de Oviedo 1959: Lib. XXIII, cap. LI, pp. 249-250):

...if any poor indian cannot fulfill what is stated as tribute, either because of sickness or poverty, or he could not find where to work, the tequitlato [tax collector] says to the lord, that such person did not want to comply with the tribute part that he was given; and the lord orders the tequitlato to take such vassal that did not want to comply, to take him to sell as a slave in a tianguez, which means a market, which is made every five to five days in all the towns of the region, and with the price with which that Indian is sold, the tribute is paid (my translation).

Craft production was clearly an important aspect of the local Tepeaca economy. Trading craft goods and providing craftsmanship[ix] were probably one of the most efficient ways to acquire food resources during periods of regional agricultural food shortages. The *Relación de Tepeaca* suggests wage labor was an economic activity present in Tepeaca during the Colonial period. It is in this context that the marketplace is especially important for the *macehualli* households. The marketplace represented much more than the acquisition of daily products and tributary items. The marketplace was the most efficient way to provision households with food products from nearby or distant regions. Having access to larger exchange webs allowed populations to obtain the needed products that were unavailable locally.

Macehualli would have hardly sought to obtain bulky goods outside their settlement, if they were available at a local marketplace. Transporting food staples back to the homestead such as maize or other grains was energetically and logistically expensive because virtually everything was carried on human backs. Moving staple grains for long-distances was unlikely because of the energy consumed by the porters in transporting them (Drennan 1984, 1985). Yet, there were people specialized in transporting goods known as *tameme* and it may be that transporting bulky goods was probably more common than we are assuming. Profit making, or the need to acquire the needed foodstuffs, would have been a salient reason for people to engage in such profession. Hirth (2010) has argued that, as energetically inefficient as it may seem, maize and chía moved regularly over 75 to 200km from Morelos, Puebla and Tlaxcala to be sold in the markets of the Basin of Mexico. Also, the *Relaciones Geográficas* indicate that groups from southern Puebla would travel up to 75km to purchase maize in eastern Puebla during times of food shortfalls.

One of the advantages of the marketplace was its ability to connect several populations into a complex commercial web. Marketplaces were not detached elements or circumscribed to a specific region. On the contrary, they were in constant contact with other areas within the *altepetl* boundaries and from other surrounding *altepemeh*. Today, several large populations within the study region have a tradition of periodic marketplaces including Tecamachalco, Acatzingo, Quecholac, Cuauhtinchan, Nopalucan, Amozoc and Teteles. It is probable that marketplaces were also an important part of the local economy during the prehispanic periodic because we know that Acatzingo, Quecholac and Tecamachalco had a *tianquiztli* (marketplace).[x] Each settlement was located within a reasonable walking distance of 10 to 30km at one another, well inside the walking distance for transporting bulky goods. Staple bulky products could have easily been exchanged between these settlements creating a regional trading web that connected settlements to a larger inter-regional web of marketplaces. Such an arrangement could provide access to food resources and other products from different environmental regions (Figure 81).

A constant interaction between neighboring marketplaces was highly possible because (1) relatively good loads of products of 23kg or more could be carried by porters and transported for relatively large distances of 50km or more (Hirth 2010), and (2) Tepeaca is located adjacent to environmental regions from which agricultural products and other goods could be imported on a regular basis. Regional topography has created a series of different micro-environments within the Llanos de San Juan, the Puebla-Tlaxcala and the Tepeaca valleys with substantial inter-annual climatic differences and rainfall variation. As was shown in Chapter 6, normal agricultural production differs considerably across these regions, and severe climatic disruptions can have differential effects within the Tepeaca valley. Maize production tends to vary according to latitude and altitude. The northern Llanos de San Juan and the western Puebla-Tlaxcala valley have better soil and precipitation conditions for agriculture resulting in more stable and higher maize yields.

The web of marketplaces was a useful buffering mechanism to lessen regional agricultural food shortages. Food could be mobilized from distant areas in times of need through the inter-regional web of marketplaces in a series of exchange relays. A problem in doing so would have been the rise of price due to the accumulated transport costs and profit gain of intermediaries. Such a scenario occurred in 1454 when the Basin of Mexico was struck by a prolonged drought that endured for three years and which caused crop failure and a massive famine among commoners (Hassig 1981). The demand for food produced such a sharp rise in value that people '…would trade a child for a small basket of maize…' (Durán 2006 [1579]: Chapter XXX, p. 243). On the contrary, the Totonac region in the coastal areas of the Gulf of Mexico had an abundance of food and brought grain to the Basin of Mexico in order to trade it for slaves:

> During this time, the Totonac people had harvested abundant grain; and when they heard of the great need of the entire land of Mexico, and how family members sold one another, they wrought a vengeance upon the Aztecs. They came to Tenochtitlan carrying great loads of maize in order to buy slaves. They also went to other cities–Texcoco, Chalco,

---

[ix] 'Y, asimismo, los naturales villanos se alquilan y sustentan de sus trabajos, que acuden a todas las obras que saben hacer. Y, destas sus granjerias, pagan lo que deben de sus tributos, ansí a su Maj[es]t[ad] como a sus encomenderos' (Molina 1985: 257).

[x] "Licencia a los indios de Acacingo para que puedan hacer el tianguez segund e como solían hacer antes" (1542, AGN, Mercedes, vol. 1, exp. 370, f.172v.).

"Para que el alcalde mayor de Tepeaca haga relación a esta Real Audiencia cerca del pro o daño que se sigue de hacer en un mesmo dia tianguez en los pueblos de Acacingo e Tecamachalco, conforme lo aquí contenido" (1566, AGN, Mecedes, vol. 8, exp. s/n, f.236r-236v.).

"Y ansí, los hortelanos lo tienen por granjería, y, ansí estas verduras de Castilla como las de la tierra, l[as] sacan a vender a las plazas y mercados. Estas verduras de Castilla se dan bien en esta tierra y provincia, especialmente en Tecamachalco, Quecholac y Acatzingo, por ser [el] agua gruesa y haber alguna abundancia" (Molina [1580] 1985: 253).

Agricultural Productivity and Tribute in 16th Century AD Tepeaca

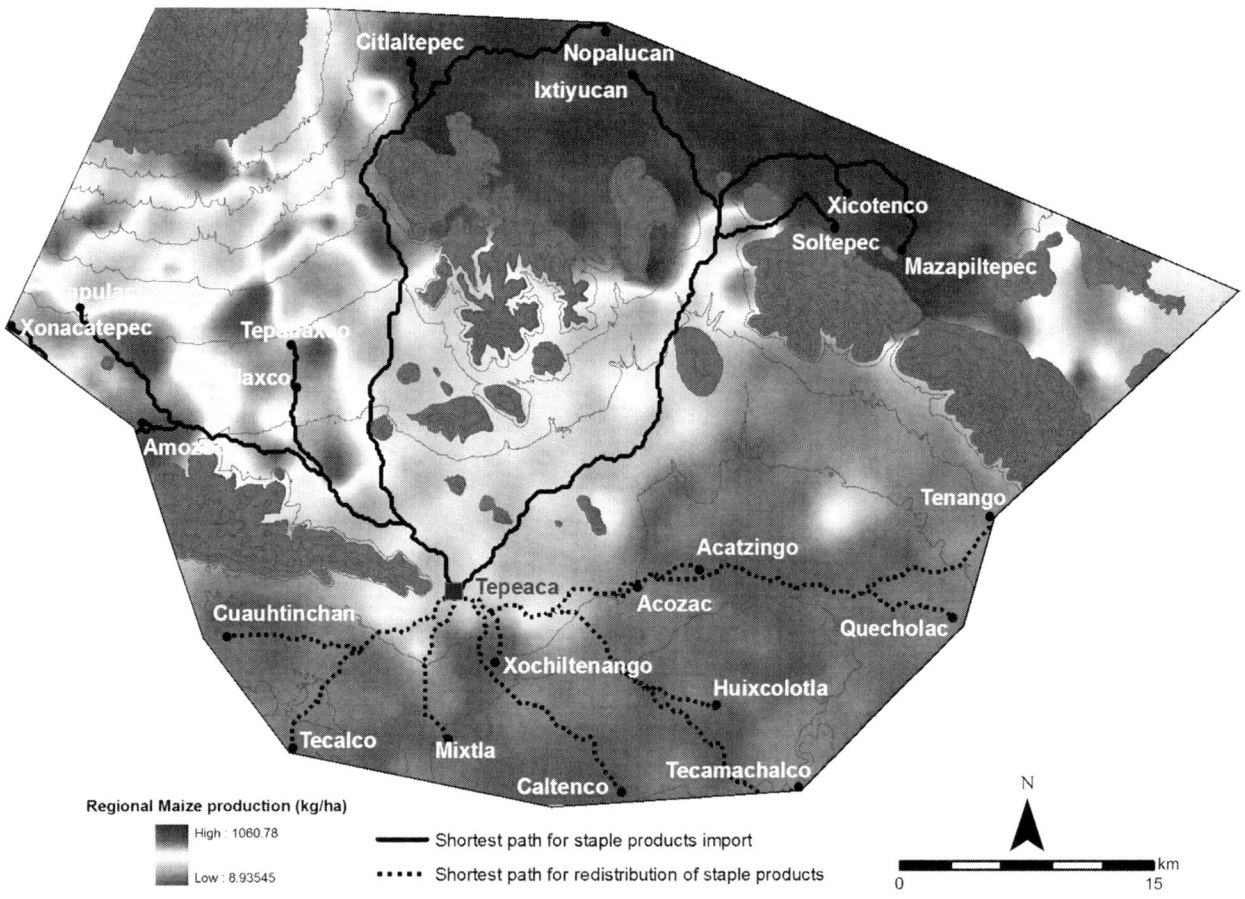

FIGURE 81. THEORETICAL LEAST COST ROUTES OF STAPLE PRODUCT MOBILIZATION FROM AREAS OF GOOD PRODUCTIVITY TO AREAS WITH FOOD SHORTAGE USING THE TEPEACA MARKETPLACE AS A PROVISIONING CENTER.

Xochimilco, and the Tepanec center, Tacuba–where they purchased large numbers of slaves with their corn. They placed yokes around the necks of adults and children. Then the slaves, lined up one behind another, were led out of the cities in a pitiful manner, the husband leaving his wife, the father his son, the grandmother her grandchild (Duran 2006 [1579]) (my translation).

Research has shown that marketplaces were key economic institutions across the Mexican highland that functioned to mobilize, convert, and distribute food and other key resources between elite and non-elite sectors of the economy (Berdan 1989; Blanton 1996; Brumfiel 1980; Carrasco 1983; Garraty 2006; Hassig 1982; Hirth 1998; Smith 1979). *Macehualli* would have used marketplaces and the regional networks they provided for obtaining food during times of shortage. Nobles probably did not need to obtain food resources from the marketplace for their own needs because they were able to amass considerable large amounts of agricultural surpluses during average years.

In contrast, *macehualli* households at some point or another probably had serious problems with annual food provisioning. Variable rainfall patterns could have generated years of markedly different agricultural production levels. In years with good rains, agricultural production could have boosted the household's overall economic well-being and stimulated the local economy. For example, *The Relación de Tepeaca*[xi] tells that the price of maize, wheat and other local legumes fell considerably in years of good precipitation. On the contrary, production declined considerably during periods of regional droughts. This inter-annual variation in production put *terrazguero* households at considerable risk. When shortages were inevitable, households had to acquire resources from other areas via exchange or reciprocity.

Macehualli/terrazguero probably took the most advantage of the inter-regional web of marketplaces in order to acquire the necessary supplies for their sustenance. This was especially important for populations practicing subsistence agriculture in this region because they faced some degree of food shortage on a reoccurring basis. Maintaining the marketplace was also important for the political infrastructure because procuring a stable supply of food to *macehualli/terrazguero* ensured the survival of the tributary base on which it rested. Without the commoner sector, local political institutions could have not been sustained nor would wealth production have been feasible.

[xi] "Y, el año que aciertan, producen las semillas de maíz y trigo, y otras legumbres de la tierra, en tanta cantidad, que hace abaratar los precios en esta provincia y su comarca" (Molina [1580] 1985: 230).

Summary

In this chapter, I examined the role of the *macehualli/ terrazguero* as the agricultural supporters of Tepeaca political institutions. I have demonstrated that agricultural production was a major means for financing the system of stratification.

The *teccalli* (*tlahtocayo*) promoted a highly polarized system of land tenure were nobles controlled vast extensions of patrimonial lands and large numbers of *macehualli/terrazguero* tributaries within their domains. *Terrazgueros* preformed a key function within the Tepeaca polity by generating the necessary agricultural food for the sustenance of the community. However, agricultural tribute imposed on *macehualli/terrazguero* as obligation involved tilling a relatively small area of around 0.17ha. This was probably the optimal strategy developed by the nobility because it gave commoners the opportunity to work their own plots and meet household consumption needs, thus ensuring their survival. However, because nobles possessed large estates distributed over wide areas, they could amass and mobilize substantial amounts of maize surplus that could reach hundreds of metric tons annually. This condition also allowed nobles to accumulate wealth and cope more efficiently with climatic variability and severe droughts.

I have also shown that variable climatic events were one of the most important restraints for Late Postclassic populations in the Tepeaca Region. The Tepeaca subsistence agriculture was limited in various ways. *Terrazgueros* cultivated relatively small plots (0.68 to 1.36ha) and worked fields predominantly dependent on rainfall cultivation. In addition, unpredictable and variable climatic conditions derived in unstable crop yields between years. However, commoner populations probably generated different levels of agricultural produce in accordance with the area they inhabited. Maize production in the north and western areas of the study region probably generated yields well above the household's annual food requirements, providing the possibility to generate surplus. On the contrary, households in the southeastern marginal region of the Tepeaca Valley were at constant risk of production shortfalls and food shortages due to poor soils, low precipitation levels, and high evo-transpiration rates. Households set in this location would have seen agricultural production levels constantly fall below yearly needs, and probably even more during pronounced harsh climatic disruptions and insect infestations.

Food deficits were probably a regular component of adaption for ancient populations of Tepeaca. In particular, marginal areas required that households develop cultural responses to buffer and cope with unpredictable environmental effects on cultivars. Due to recurrent food deficits, households probably needed to regularly acquire food via other resources. In areas with low agricultural productivity, households may have opted to engage in other economic activities to compensate food shortages. Domestic craft production was probably one of the most efficient ways to compensate for food deficits. Crafting activities were a main activity of *terrazguero* households and items produced were widely used as tributary payments to elite. Interestingly, having to work their small field probably allowed *macehualli/terrazguero* to work in alternative economic strategies such as multi-crafting and intermittent craft production. Trading such craft goods were probably an efficient way to acquire food resources during periods of regional agricultural food shortages. I believe that the local and periodic marketplaces were especially important for *macehualli* households. The marketplace represented a way to acquire daily products, tributary items, and food products from nearby or distant regions through a web of regional marketplaces.

Chapter 9
Conclusions and Directions for Future Research

In this chapter I present the conclusions derived from this research. The general goal of this study has been to analyze the impact of climatic variability and human managing strategies on Late Postclassic (AD 1325-1521) and Early Colonial (16th century AD) agricultural systems of the Tepeaca Region. This work centered on the following three main goals: (1) model agricultural productivity at the household and regional levels, (2) identify the buffering strategies developed by Tepeaca's populations against cyclical food shortfalls, and (3) establish a model for the agricultural and economic structure of the Tepeaca *altepetl* or state-level polity.

Modeling agricultural productivity in ancient Tepeaca

A central theme of this study has been to model agricultural productivity in ancient Tepeaca at the household and regional levels. I have done so by analyzing the effects of climatic variability on maize production using an analogical approach. The logic followed is that similar environmental and human constraints on agricultural productivity affected past populations as it does today (Halstead and O'Shea 1989). Studying the relationship between climatic variability and agricultural systems is crucial for establishing possible human adaptations to unpredictable environmental factors. It also serves as a mechanism for evaluating the impact of regional and local food shortages on the economic structure of prehispanic societies.

Rainfall agriculture was probably the main form of agriculture practiced in Tepeaca during the Late Postclassic and Early Colonial periods. This type of agriculture is particularly susceptible to shifts in rain patterns and other negative climatic phenomenon. Yields are highly variable between years, making it a risky and unstable form of food production. Despite this fact, in most cases researchers have preferred to employ average yield values for estimating the production capacity of ancient agricultural systems. Averages are a useful way to approach the internal economic structure of political entities. Yet, the use of averages can obscure some of the 'normal' variations in food deficit at the local and regional levels. For this reason, this study has taken advantage of using variable production estimations in addition to averages when reconstructing Tepeaca's indigenous agricultural economy. This approach allows implications to be made regarding staple production at the household level and the capacity of Tepeaca's political institutions to generate surplus.

In this work I have approached ancient indigenous agricultural productivity by analyzing the factors that shape modern maize production within the study region. Throughout this work, I have stressed out that maize production is intrinsically volatile when carried out under rainfall conditions. This condition is embodied in the variable and sometimes contrasting historical maize yield averages registered within the study region. Historical maize yields in Tepeaca average 1430kg/ha. This value is low compared to maize production under other intensive forms of agriculture like irrigation and terracing (Armillas 1971; Patrick 1977; Sluyter 1994), but well within the worldwide average yields. Nevertheless, a closer look into yield values shows that maize productivity can be extremely variable between years. Yields range from a few kilos per hectare to more than 3500kg/ha depending on the local environmental conditions and the year's climatic patterns in which maize was cultivated.

As a main component of this research, in 2009 I conducted an ethnographic survey of 49 fields from 24 households within the study region aimed at analyzing the effects of climatic variability at the regional and household levels. The ethnographic survey coincided with a major canícula seasonal climatic disruption that was probably intensified by the 2009-2010 El Niño events. Such combination of factors derived in a severe regional drought. The drought resulted in regional crop failure that struck differentially within the study region. This unfortunate event for local farmers became an opportunity for measuring the degree of effect on maize yields caused by this phenomenon. For this reason, I arbitrarily decided to expand my sample to 500 fields encompassing different environmental areas within the 1300km² study region. The goal was to establish the prominent variables (e.g., soil properties, moisture retention in the subsoil, human buffering strategies) affecting maize production when rainfall shortages occur during the dry canícula and the regional distribution of its effects on agricultural crops.

Gathering data on maize yields from 500 fields conferred several problems. The main difficulty was the need to develop a sampling method that permitted me to estimate maize yields over large areas in a relatively short span of time. To do so, I developed the Ear Volume Method (EVM) for estimating maize yields. The EVM estimates yields by calculating the average volume of the ear from a sample of plants. Ear volume can then be converted into kilograms of shelled maize produced by plants. The average weight of shelled maize by plant could then be multiplied by the estimated plant densities of each field, thus obtaining the kilograms of shelled maize produced per hectare. The EVM implies that there is a significant relationship between ear dimensions and its equivalent in weight of dry shelled maize. Such correlation was tested successfully using a bivariate correlation and a regression analysis. The EVM has the advantage that the researcher can make relatively fast in-field estimations of yields with a reasonable degree of precision. The procedure is time efficient because only three simple measures need to be taken from ears without having to rip the ear from the plant. In addition, maize yields can be estimated before the ears are harvested. There is no need to worry about moisture content in the ear because the EVM method seeks to obtain volume not weight.

A difficulty with the EVM is that it can only be carried out when maize ears have reached their full maturity. Therefore, one has to wait between four to six months to be able to measure ear volume in maize plants, depending on the time it takes local landraces to fully develop. It is important to emphasize that neither ear volume nor cob length can by themselves give a reliable estimate of maize yields. In order to achieve a good estimate the researcher needs to estimate total plant densities per unit of space. Furthermore, there may be the need to consider maize prolificacy as an important variable. In this work, I have considered that maize plants generally produce a single cob when cultivated in rainfall agriculture. However, other Artificial Water Supply agricultural systems (AWA) such as irrigation may produce prolific plants with more than one ear per plant as a consequence of better moisture and nutrient conditions for plant development. In this case, the researcher may consider it more appropriate to estimate yields by multiplying the average weight of shelled maize per ear by the estimated number of ears per unit of space.

The results of the 2009 data analysis indicate that droughts differentially affected each geographical zone to varying degrees. This work has demonstrated that the most favorable areas for maize cultivation are located in the northwest portion of the study region within the Puebla-Tlaxcala valley and the Llanos de San Juan. These areas have lands of Class 1 and 2 in which maize averages between 2222kg/ha and 2459kg/ha. Lands in the northwestern portion of the study region like Amozoc, Acajete and Zitlaltepec probably had the potential to generate maize above that needed to sustain a typical household of 7 members annually (Sanders, et al. 1979: 372-373, Table 1; Steggerda 1941). On the contrary, the south and southeastern portions of the Tepeaca Valley, including Caltenco, Tochtepec and Tecamachalco, probably had the worst conditions in which to carry out agriculture. Today, these marginal areas have the lowest maize yields within the study region, which in most years produce averages below 350kg/ha.

The 2009 dry canícula data provides important information regarding the repercussions of severe climatic disruptions over regional agricultural production. The canícula is a well-known phenomenon among agriculturalists, but its degree of effect can vary between years. Events such as the 2009 drought were probably unusual and very hard to predict in ancient times as they are today. There is some information that suggests large scale cyclical climatic shifts such as ENSO may influence the severity of the canícula period; yet, the degree of impact can be variable in time and space with no apparent trend (Appendini and Liverman 1994; Peralta et al. 2008). Such uncertainty in rainfall precipitation would have certainly created a sense of awareness among the Tepeaca prehispanic populations. It would also lead them to prepare against the possible occurrence of a regional food shortage.

By analyzing the 2009 regional drought on maize production we can deduce some of its implications for human adaptive strategies to cyclical food shortages. It would be unwise to believe that the regional drought generated by the 2009 canícula would have occurred in the exact same way in past times. Yet, the data allows us to make some important annotations. The most evident is that in prehispanic times a severe canícula would have generated a widespread crop failure that affected more deeply those areas located in the south portion of the Tepeaca valley. Such areas include towns like Acatzingo, Caltenco, Tecali, Cuauhtinchan, Tochtepec and Tecamachalco where Class 3 to 5 lands predominate, many of them with extreme marginal conditions for agriculture. In the southern zones with Class 5 lands, droughts probably intensified the harsh conditions causing maize cultivation to fluctuate at levels below 350kg/ha. Oddly, the poorest productive Class 5 lands in Tecamachalco recorded the lowest percent of anomaly (-30.92%) in yields in 2009 (see Figure 66). This can be explained considering that yields in this area are extremely low. Consequently, in ancient times production decreases resulting from a major drought would have represented a relatively minor blow to food availability for households settled in this zone. With or without a severe drought, populations in the south and southeastern zones of the Tepeaca valley had very low levels of maize production that probably forced them to obtain the necessary resources for their survival by other means.

Another implication derived from this research is that severe climatic variability was probably not homogenous throughout the Tepeaca region. While some portions may have suffered profound levels of crop failure others probably were affected at a lesser degree. The 2009 canícula caused severe damage to cultivars in the Tepeaca valley, but its effects were much lesser in the adjacent Llanos de San Juan. This is important, because it implies that food resources could be mobilized from the areas that produced well to those that did not through reciprocity or exchange networks.

Identifying buffering strategies against cyclical food shortfall

The 2009 ethnographic survey showed that unusual climatic fluctuations such as the canícula had important repercussions for ancient populations, the most salient one a widespread decrease in food production. Therefore, a second goal of this study was to identify some of the possible buffering strategies developed by populations in order to cope with cyclical food shortfalls. Research has shown many ways in which farmers can mitigate the negative effects of climate over crops, many of them dealing with small scale developments or adaptations (Wilken 1987). The five buffering strategies that ancient indigenous populations in Tepeaca could have employed in order to lessen the risk of total crop failure were (1) field dispersion, (2) differential sowing/harvest times throughout the region, (3) the use of intensive cultivation techniques, (4) the adoption of alternative economic activities aside agriculture, and (5) the use of alternative food resources.

Field dispersion is a widespread strategy used by farmers aimed at diversifying risk of total crop failure due to environmental unpredictability. Locating agricultural fields in different areas within 4 to 5km from the household serves to avoid losing the entire crop and to average overall yields. The 2009 ethnographic data revealed that locating fields strategically in various environmental areas might not necessarily lessen the negative effects of a major climatic disruption. Events such as the 2009 dry canícula generally affect large areas that easily surpass the walking distance at which fields are located from the household compound. Under these circumstances, the lack of precipitation makes field dispersion an insufficient buffering mechanism against major droughts.

This has important implications for Late Postclassic and Early Colonial smallholders from the Tepeaca region. Field dispersion might be a useful strategy to fight back minor shifts in rain patterns including the typical midsummer lull of around one month. However, *macehualli/terrazguero* households would have had little chance to fight back a major disruption in precipitation patterns. On the contrary, the indigenous Tepeaca elite were in a much better position to buffer total crop failure caused by regional droughts. Nobles also used field dispersion as a buffering strategy by controlling lands in several parts within the Tepeaca *altepetl*, many of them located in different environmental regions. By using a much wider space, some fields of the elite would eventually avoid the negative effects of severe climatic events. The case of the local noble Doña Francisca de la Cruz, presented in Chapter 8, exemplifies the advantages of averaging total maize production from several fields in many locations. Doña Francisca's *teccalli* probably produced thousands of metric tons of maize annually. Such levels of production would have allowed noble households to store high amounts of food resources during average years, allowing them to be in a good position to buffer any eventual food shortage. Their agricultural potential was so high that even under a severe drought they could have generated sufficient maize to satisfy their annual caloric requirements.

The second buffering strategy was to have fields sowed/harvested at different time intervals throughout the region. As stated in Chapter 6, local environmental factors shape the start and end of the agricultural cycle throughout the Tepeaca region. Farmers sow at different moments due to several factors including the need to take advantage of the early rains, avoid early frosts, and organize agricultural tasks in various fields. Additionally, individual decision-making and the use of a variety of different local maize landraces produce a mosaic of initial sowing taking place at different times. This ultimately derives in harvests taking place on a staggered schedule throughout the region. It also confers a certain degree of security that a proportion of cultivars will avoid climatic reversals and produce a given quantity of food. For example, maize cultivation in the western portion of the Llanos de San Juan region was less affected by the 2009 drought probably because (1) the lack of water did not coincide with the crucial ear development in maize, and (2) frosts fell after the maize ears were fully developed and could not be affected by the low temperatures. In prehispanic and Early Colonial times, sowing and harvesting at different time intervals could have been an important strategy for food provisioning at the local and regional levels, especially for the political sectors that had fields located in different areas within the Tepeaca *altepetl*. Differential sowing/harvest patterns could allow elite and commoners to have a constant supply of maize coming into local warehouses and marketplaces from as early as September to as late as January. Such a pattern would have helped insure a constant supply of food provisions throughout most part of the year.

A third buffering strategy could have been to carry out intensive cultivation techniques. An interesting find of this work is that although the 2009 canícula drought caused a 30 to 90% decrease in maize yields throughout the study region, in no case did I registered total crop failure on a field (a value of zero). Even in the worst scenario farmers were able to harvest a minimal amount of maize. This suggests that prehispanic populations could have lessened some of the negative effects of droughts on plants by simply carrying out more intensive plant rearing techniques such as plant transplanting, pot irrigation, and manual ear prolificacy suppression in maize (see Chapter 5). These techniques probably promoted a more efficient use of the available resources and helped mitigate local environmental constraints given the available technology.

A fourth buffering strategy may have been to adopt alternative economic activities, such as domestic crafting. Such activities are aimed at provisioning the household with food resources under times of regional food shortages. If cultivation provided enough food for the annual subsistence needs of households, then small parcel cultivation could enabled them to pursue additional economic activities. In particular, multi-crafting and intermittent craft production were important income sources for increasing household economic well-being (Hirth 2009). Local historical sources clearly show that alternative economic activities were common in Late Postclassic and Early Colonial Tepeaca and included a wide array of professions (see Figure 80). Alternative subsistence activities besides agriculture were important for mobilizing food resources and to cover the expenses and needs of daily life. Trading craft goods and providing craftsmanship were probably one of the most efficient ways to acquire food resources during food shortages.

In central Mesoamerica, including the Tepeaca region, local and periodic marketplaces called *tianquiztli* were a main feature of indigenous societies. Marketplaces were key economic institutions across the Mexican highland that functioned to mobilize, convert, and distribute food and other key resources between elite and non-elite sectors of the economy (e.g., Berdan 1989; Blanton 1996; Brumfiel 1980; Carrasco 1983; Garraty 2006; Hassig 1982; Hirth 1998; Smith 1979). Marketplaces were the most efficient way to provision households with food products from nearby or distant regions. Having access to larger exchange webs allowed populations to obtain the needed products unavailable locally. This aspect would have been especially important to *macehualli* populations as mechanism for obtaining food during times of shortage. The marketplace as a regional food supplier is compatible with the second buffering strategy of differential sowing/harvest patterns. Food produced at different time intervals throughout the region, or between regions, could be mobilized from distant areas in times of need through the inter-regional web of marketplaces in a series of exchange relays.

A fifth buffering strategy against food shortfalls may have been the use of alternative food resources aside from maize. Although hypothetical, it is probable that households living in marginal environment conditions like the southern portion of the study region could have opted to make use of other locally available food resources. The semi-arid environment of the south strip of the Tepeaca valley promotes the appearance of a different floral and faunal biota that is not found on the higher altitude northern lands. It may be that households in the south exploited local plant and animal species similar to those reported by Feinman and his colleagues (2007) in El Palmillo, Oaxaca. Some of the useful plants and animals with an important caloric source include maguey, izote trees, seasonal fruits like *tuna* (prickly pear) and *pitaya* (dragon fruit), and various animals like hares, rabbits and deer. As Anderson (2009) has pointed out, maguey cultivation in particular may have been an important caloric source for households with marginal lands not adequate for growing cereals such as maize. However, at this point we can only make inferences in this respect. Further research may address this issue more adequately.

The dual agricultural economic structure of the Tepeaca *altepetl*

The third goal of this study was to establish a model for the agricultural and economic structure of the Tepeaca *altepetl*. The model developed in this work implies that ancient Tepeaca agricultural systems were arranged dually as subsistence and institutional agriculture. This view is highly useful for it provides both a 'top-down' and 'bottom-up' perspective for studying the agricultural economic structure of Late Postclassic and Early Colonial populations in central Mesoamerica. Studies that focus on the role of political apparatus have provided key information on the internal structure of indigenous complex societies. However, studies that focus on the commoner sectors are also becoming increasingly central in Mesoamerican archaeology, allowing us to achieve a more comprehensive picture of the past (e.g., Joyce 2001; Lohse and Gonlin 2007; Restall 2005; Restall *et al.* 2005; Robin 2003; Smith and Heath-Smith 2000; Webster 1988). This work has employed such a perspective by proposing a model that considers an interdependent a dual system of agricultural production in which the focus is the indigenous community and not only one sector of it.

In ancient Tepeaca, institutional agriculture dealt largely with production for the support of elite and financing political institutions. In Early Colonial times, it was limited to support the local nobility, but in prehispanic times it included a broader spectrum of military, theocratic, and political institutions. The Tepeaca political institutions required both staples and labor to finance socio-political stratification. The objective of the local political systems was to maximize income and to grow in proportion to their access and control over the key resources. For the Tepeaca nobility, agricultural production was an important component of the tributary system and a key element in wealth accumulation. Consequently, the nobility of Tepeaca used agricultural production at the household level to support indigenous social institutions. The accumulation of wealth through agriculture was certainly unpredictable because fluctuating environmental conditions provided a constant threat of crop failure. Yet, agricultural production with other tributary sources was successfully able to generate wealth.

Running alongside institutional agriculture was subsistence agriculture. Subsistence agriculture was performed by the commoner *macehualli* sector and characterized by a relative low-level production capacity geared towards fulfilling auto-consumption needs of the household. For *macehualli* and the landless *terrazgueros*, agricultural production was also unstable. The most important factors determining the level of food production among commoner households were variable climatic phenomenon, the local environmental conditions in which to perform agriculture, and the size of agricultural fields. The combination of these three variables determined the household's capacity to generate surplus or to suffer a certain degree of food shortage during its life. In most portions of the study region, maize production was sufficient to provide the household with the annual caloric requirements of its members and provide an additional surplus for storage.

In ancient Tepeaca subsistence level agriculture allowed the household to survive and still be able to make a substantial contribution to the tributary obligations. The Tepeaca *teccalli* (*tlahtocayo*) economic and political apparatus was able to propagate itself thanks to the commoner base that tended their fields and produced wealth items that intensified social hierarchies. In addition, the *terrazguero* sector became the main basis for tribute extraction in the Early Colonial period. The payment of agricultural tribute was well-regulated. It included the allotment of standard areas of 600 squared *braza* or *cemmatl* (1673m^2) to the tributary *terrazgueros* who rented lands from nobles. For individuals with other occupations such as construction workers and merchants, the agricultural tribute area was much lower, probably below 240 squared *braza* or *cemmatl* (669m^2). In the prehispanic period, other *macehualli* sectors with patrimonial lands also cultivated a section of land that was designated for institutional use. The area cultivated collectively by *macehualli* covered 400 square *nehuitzantli* (1743m^2). In all cases, these allocations represented manageable areas for tribute cultivation purposes from the standpoint of energy expenditure. It allowed commoner households to work their own parcels intended for self-consumption, while complying with other tribute demands such as personal service and the manufacture of cotton textiles.

Given that most *macehualli* households from Tepeaca were probably self-sufficient and had the capacity to generate some amount of surplus has important anthropological implications for the study of indigenous Mesoamerican economic systems. In spite of the tributary burden imposed on commoner populations, it is probable that the Tepeaca elite did not interfere with food production at the domestic level. Rarely would they have alienated key resources for the survival of the commoner base. Such perception was clearly exemplified in the landlord-tenant arrangement that existed between the Colonial *terrazgueros* and tlahtoque. Land was clearly separated. On one side there were a number of fields (4 to 8) allotted to landless *terrazgueros*. They were destined strictly for the sustenance of *terrazguero* households. On the other, each *terrazguero* household had to work an additional field as the tributary payment for the right to use the noble's lands. If nobles had not honored such an arrangement, they would have undermined the ability of *macehualli* to cope with unstable climatic phenomena and food shortage events. It was in the interest of elite to keep such an order because it assured them with a tributary payment that could perpetuate the *teccalli* system and avoid social insurgency.

Another important element of subsistence agriculture in prehispanic and Early Colonial Tepeaca is that it was probably highly unstable and unpredictable. The Tepeaca *macehualli* households were susceptible to seasonal climatic disruptions and severe drought events produced by strong climatic disruptions. The dry canícula, and other climate perturbations associated with the ENSO phenomenon (Florescano 1980), can occur at any point and indigenous populations must have been well aware that such events would eventually arrive. However, what we learn from modern occurrences such as the case of the 2009 regional drought is that adjacent environmental regions respond differently to any major climatic disruption. Elevation is a crucial factor producing differential climatic and precipitation regimes, which is reflected directly in levels of agricultural productivity. In Late Postclassic and early Colonial times, such differences had to create significant differences in food availability among indigenous settlements. Crop failure must have been a constant preoccupation for indigenous populations, but probably it differed considerably between years and different households. While in one region, agricultural production was bad, in others it performed regularly or good.

As mentioned before, the most severe effect of severe droughts had to occur in the semi-arid southern areas of the Tepeaca Valley where environmental and climatic conditions are more marginal. The areas in the northern Llanos de San Juan and the western Puebla-Tlaxcala valley were much more stable and productive. Not surprisingly, controlling the northern zones was one of the most important military strategies undertaken by Tepeaca during the Late Postclassic. By controlling these territories, the Tepeaca polity was able to maintain a steady agricultural input by way of tributary impositions on local *macehualli*. This also allowed the local elite to settle in the Tepeaca area in order to control the key commercial route that passed through the Valley of Tepeaca as well as the local marketplace (Martínez 1994b).

Maize production capacity at the institutional agriculture was also unstable. Yet, elite nobles were capable of generating large amounts of surplus above the needs of noble household members. This allowed them to finance other political institutions that were not part of the agricultural labor force of the *altepetl*. Such strategy was possible because nobles controlled large territories, some of which were probably outside the altepetl's core area. Having control over the land allowed nobles to extract wealth through human labor via the terrazgo, service and crafting activities. The elevated amounts of agricultural tribute mobilized by nobles allowed them to efficiently buffer any localized seasonal climatic fluctuation and crop failure. It is possible that even under severe crop failure, nobles would still be able to amass a substantial amount of surplus.

The unequal distribution of land and the variable nature of agricultural production generated a pattern in which the risk of total crop failure was concentrated within the household realm. *Terrazgueros* were smallholders with their fields concentrated in relatively small areas. This pattern of land use made them susceptible to highly variable annual production losses due to localized and regional climatic fluctuations. Elites generated greater quantities of staple products because their large landholdings were spread over larger areas, thus avoiding climatic variability and crop failure much more efficiently. Land control also allowed nobles to also control landless peasant labor and tribute. This mainly took the form of payment of wealth items and personal services rather than on their agricultural food base.

Directions for future research

The 2009 regional maize productivity survey provided an important corpse of data concerning the impact of severe climatic phenomenon on agricultural production at the local and regional levels. However, the information collected pertains to a single year's maize production at the regional level. Most importantly, the data recorded the effects caused by an unusual climatic event that may not be representative of a typical year's agricultural productivity. For this reason, it is imperative that future research be directed at carrying out similar surveys to record maize production under a typical year, and preferably for a series of years. Such data will important in order to compare information with that of the 2009 season.

From the 2009 maize production survey one crucial aspect has arisen with respect to the southern strip of the Tepeaca valley. Maize production in this area is extremely low due to the local marginal conditions, especially the low precipitation regimes and the poor quality soils. Two implications can be derived that need further study. On one hand, maize may not have been the main staple grown in these areas. Environmental conditions are unfavorable for maize plants to grow adequately due to poor soils, time restricted precipitation periods, and high evo-transpiration rates. Future research should be directed at obtaining information on ancient plant use among these households. It is possible that alternative species more tolerant to moisture deficits were exploited, like those reported in similar marginal conditions of El Palmillo, Oaxaca (Feinman, *et al.* 2007). On the other, we must consider the possibility that climatic conditions were different during the Late Postclassic and Early Colonial periods than they are today. If seasonal precipitation was higher, then it is possible that the marginal southeastern region had an adequate environment for maize cultivation. It may also be that rain patterns vary cyclically alternating between periods of good and bad precipitation. This may be a possibility considering that modern farmers say that 30 or 40 years ago the region received much more precipitation and had a cooler climate than it does today. In addition, people recurrently speak of the presence of several springs located at different points of the Tepeaca valley. I have tried to locate them, but it appears that recent extraction of water well has exhausted local aquifers and springs are no longer active (Flores *et al.* 2006). If local springs were available in prehispanic times, this would imply the possibility of having better conditions for maize cultivation using intensive technologies such as canal or pot irrigation. Therefore, it is imperative that future studies be directed at answering this important question.

It is crucial that future research also be directed at better understanding seasonal variation in precipitation in the archaeological record. As I have shown, archaeologists should not rely only on average precipitation values because they are not necessarily a good indicator of a good or bad year's harvest. Receiving the necessary amount of water is crucial for plant cultivation. However, most important is the time when water is applied. In maize, plant development, especially ear growth, will be poor if water does not arrive at the right moment. As far as I know, we have not yet identified prehistoric severe seasonal droughts in the Puebla-Tlaxcala region such as the 2009 canícula. A possible way to detect finer inter-annual climatic anomalies in the Puebla-Tlaxcala region would be to carry out analysis of ancient lacustrine environments and deep alluvial soils such as the works of Borejsza *et al.* (2008), Lauer (1979), Ohngemach and Straka (1978), Rivera *et al.* (2007), and Xelhuantzi (1994). Also, the use of tree ring analysis is another possibility, as has been done in other parts of Mexico (Stahle *et al.* 2011). Given that tree-ring development is strongly influenced by temperature, rain patterns, and natural phenomenon (e.g., Akachuku 1991; Angeles 1990; Bernarz and Ptak 1990; Dean *et al.* 1996; Hughes *et al.* 2009; Putz and Taylor 1996; Yamamoto 1992), ancient atypical climatic events such as severe droughts could be detected by analyzing wood ring composition and morphology. The effects of climate on radial growth may be more evident on some tree species than others. However, a likely candidate to perform such studies could be *ahuehuete* trees (*Taxodium macronatum*). This specie can be found in many parts of the Puebla and Tlaxcala regions and are known to live up to hundreds of years. Tree-ring data have been used to model ancient and modern crop failure scenarios (Anderson *et al.* 1995; Burns 1983; Cleveland *et al.* 2003; Stahle and Cleveland 1994; Therrell 2005). Such information would be extremely useful for identifying ancient shifts in precipitation and climate and would provide us with better elements to reconstruct possible crop failure catastrophes within prehispanic societies.

Finally, future research needs to evaluate if the Ear Volume Method for estimating maize productivity can be applied to other regions. The EVM was most useful in the Tepeaca region. I believe that it can also be employed in other regions of the central Mexican highlands that cultivate conical native landraces of maize. However, its reliability for cylindrical lowland landraces remains uncertain, such as those found in the Oaxaca, Veracruz and the Yucatan peninsula regions.

References

Abrams, E. M. 1995. A Model of Fluctuating Labor Value and the Establishment of State Power: An Application to the Prehispanic Maya. *Latin American Antiquity* 6: 196-213.

Aceves, E., A. Turrent, J. I. Cortés, and V. Volke. 2002. Comportamiento agronómico del híbrido H-137 y materiales criollos de maíz en el Valle de Puebla. *Revista Fitotecnia Mexicana* 25: 339-347.

Adams, K. R., D. A. Muenchrath, and D. M. Schwindt. 1999. Moisture Effects on the Morphology of Ears, Cobs and Kernels of a South-western U.S. Maize (*Zea mays L.*) Cultivar, and Implications for the interpretation of Archaeological Maize. *Journal of Archaeological Science* 26: 483-496.

Adams, R. M., L. L. Houston, B. A. McCarl, M. Tiscareño L., J. Matus G., and R. F. Weiher. 2003. The benefits to Mexican agriculture of an El Niño-Southern Oscillation (ENSO) early warning system. *Agricultural and Forest Meteorology* 115: 183-194.

Aguilar, M. 2006. *Handbook to Life in the Aztec World.* New York, Facts On File, Inc.

Akachuku, A. E. 1991. Wood growth determined from growth ring analysis in red pine (Pinus resinosa) trees forced to lean by a hurricane. *AIWA Bulletin* 12: 263-274.

Albores, B., and B. Johanna. 1997. *Graniceros: Cosmovisión y Meteorología indígenas en Mesoamérica.* México, D.F., El Colegio Mexiquense.

Almazan, M. A. 1999. The Aztecs States-Society: Roots of Civil Society and Social Capital. *Annals of the American Academy of Political and Social Science* 565: 162-175.

Altieri, M. A. 1987. *Agroecology: The Scientific Basis of Alternative Agriculture.* Boulder, Westview Press.

—. 1990. Why Study Traditional Agriculture? in C. R. Carroll, J. H. Vandermeer, and P. M. Rosset (eds.), *Agroecology*: 551-564. New York, McGraw-Hill Publishing Company.

Alva Ixtlilxochitl, F. d. 1997. *Historia de la Nación Chichimeca.* Vol. I y II. México, D.F., Universidad Nacional Autónoma de México.

Alvarado Tezozomoc, F. 1944. *Crónica Mexicana.* México, D.F., Leyenda.

Anderson, D. G., D. W. Stahle, and M. K. Cleaveland. 1995. Paleoclimate and the Potential Food Reserves of Mississippian Societies: A Case Study from the Savannah River Valley. *American Antiquity* 60: 258-286.

Anderson, J. H. 2009. *Prehispanic Settlement Patterns and Agricultural Production in Tepeaca, Puebla, Mexico, A.D. 200-1519.* Unpublished PhD thesis, The Pennsylvania State University.

Angeles, G. 1990. Hyperhydric Tissue Formation in Flooded Populus tremuloides Seedlings. *IAWA Bulletin* 11: 85-96.

Anguiano, M., and M. Chapa. 1982. Estratificación social en Tlaxcala durante el siglo XVI. In P. Carrasco and J. Broda (eds.), *La estratificación social en la Mesoamérica prehispánica*: 118-156. México, D.F., Instituto Nacional de Antropología e Historia.

Appendini, K., and D. M. Liverman. 1994. Agricultural policy, climatic change and food security in Mexico. *Food Policy* 19: 149-164.

Armillas, P. 1971. Gardens on Swamps. *Science* 174: 653-661.

Arnason, J. T., J. D. Lambert, J. Gale, J. Cal, and H. Vernon. 1982. Decline of Soil Fertility Due to Intensification of Land Use by Shifting Agriculturists in Belize, Central America. *Agro-Ecosystems* 8: 27-37.

Báez, A., E. Ascencio, C. Prat, and A. Márquez. 1997. Análisis del comportamiento de cultivos en tepetate t3 incorporado a la agricultura de temporal, Texcoco (México). *Memorias del III Simposio Internacional sobre Suelos volcinicos endurecidos*: 296-310.

Barlett, P. 1975. *Agricultural Change in Paso: The Structure of Decision Making in a Costa Rica Peasant Community.* Unpublished Ph.D. dissertation, Department of Anthropology. New York, Columbia University.

Beach, T. 1998. Soil Constraints on Northwest Yucatán: Pedo-archaeology and Maya Subsistence at Chunchucmil. *Geoarchaeology* 13: 759-791.

Beets, W. C. 1990. *Raising and Sustaining Productivity of Smallholder Farming Systems in the Tropics: A Handbook of Sustainable Agricultural Development.* Alkmaar, AgBé Publishing.

Bellon, M. R. 1991. The Ethnoecology of Maize Variety Management: A Case Study from Mexico. *Human Ecology* 19: 389-418.

Benz, B. F. 1988. Clasificación y evolución del maíz mexicano. In L. Manzanilla (ed.) *Coloquio V. Gordon Childe: estudios sobre la revolución Neolítica y la revolución urbana*: 133-148. México, D.F., Universidad Nacional Autónoma de México.

Berdan, F. F. 1989. Trade and Markets in Precapitalist States. In S. Plattner (ed.): 78-107. Standford, Stanford University Press.

—. 1994. Economic Alternatives under Imperial Rule: The Eastern Aztec Empire. In M. G. Hodge and M. E. Smith (eds.), *Economies and Polities in the Aztec Realm*: 291-313. Albany, Institute for Mesoamerican Studies.

Berdan, F. F., and P. R. Anawalt. 1992. *The Codex Mendoza*. Berkeley, University of California Press.

Berdan, F. F., R. E. Blanton, E. H. Boone, M. G. Hodge, M. E. Smith, and E. Umberger. 1996. Aztec Imperial Strategies. Washington D.C., Dumbarton Oaks.

Bernarz, Z., and J. Ptak. 1990. The Influence of Temperature and Precipitation on Ring Widths of Oak (*Quercus robur* L.) in the Niepolomice Forest near Cracow, Southern Poland. *Tree Ring Bulletin* 50: 1-10.

Billman, B. R. 2002. Irrigation and the Origins of the Southern Moche State on the North Coast of Peru. *Latin American Antiquity* 13: 371-400.

Blanton, R. E. 1996. The Basin of Mexico Market System and the Growth of Empire. In F. F. Berdan, R. E. Blanton, E. H. Boone, M. G. Hodge, and M. E. Smith (eds.), *Aztec Imperial Strategies*: 47-84. Washington, D.C., Dumbarton Oaks.

Blyn, G. 1966. *Agricultural Trends in India, 1891-1947: Output, Availability, and Productivity*. Philadelphia, University of Pennsylvania Press.

Borejsza, A., I. Rodríguez López, C. D. Frederick, and M. D. Bateman. 2008. Agricultural slope management and soil erosion at La Laguna, Tlaxcala, Mexico. *Journal of Archaeological Science* 35: 1854-1866.

Boserup, E. 1965. *The Conditions of Agricultural Growth*. Chicago, Aldine.

Brito, B. 2008. *Códice Chavero de Huexotzingo: Proceso a sus oficiales de república*. México D.F., Instituto Nacional de Antropología e Historia.

Brookfield, H. 1972. Intensification and Disintensification in Pacific Agriculture. *Pacific Viewpoint* 13: 30-48.

—. 2001. Intensification, and Alternative Approaches to Agricultural Change. *Asia Pacific Viewpoint* 42: 181-192.

Brumfiel, E. 1980. Specialization, Market Exchange, and the Aztec State: A View From Huexotla. *Current Anthropology* 21: 459-478.

Brumfiel, E., and T. K. Earle. 1987. Specialization, Exchange and Complex Societies: An Introduction. In E. Brumfiel and T. K. Earle (eds.), *Specialization, Exchange and Complex Societies*, vol. 1-9. Cambridge, Cambridge University Press.

Burns, B. T. 1983. *Simulated Anasazi Storage Behavior Using Crop Yields Reconstructed from Tree Rings: A.D. 652-1968*. Unpublished PhD dissertation, University of Arizona.

Calnek, E. E. 1992. Patrón de Asentamiento y Agricultura de Chinampas en Tenochtitlán. In C. J. González (ed.) *Chinampas Prehispánicas*: 155-178. México, D.F., INAH.

Calvo, T. 1973. *Acatzingo: demografía de una parroquia mexicana*. México, D.F., Instituto Nacional de Antropología e Historia.

Carballo, D., P. Roscoe, and G. M. Feinman. 2012. Cooperation and Collective Action in the Cultural Evolution of Complex Societies. *Journal of Archaeological Method and Theory* DOI 10.1007/s10816-012-9147-2: 1-36.

Carrasco, P. 1963. Tierras de dos indios nobles. *Tlalocan* IV: 97-119.

—. 1969. Más documentos sobre Tepeaca. *Tlalocan* VI: 1-37.

—. 1973. Los documentos sobre las tierras de los indios nobles de Tepeaca en el siglo XVI. *comunicaciones* 7: 89-91.

—. 1983. Some Theoretical Considerations About the Role of the Market in Ancient Mexico. In S. Ortíz (ed.) *Economic Anthropology: Topics and Theories*: 67-82. Lanham, University Press of America.

—. 1989. Los mayeques. *Historia Mexicana* 39: 123-166.

—. 1997. La Procedencia de los datos de Zorita sobre la organización social prehispánica. In E. de la Lama and M. E. Landa (eds.), *Simposium internacional de investigación de Huexotzingo*: 85-94. México, D.F., Instituto Nacional de Antropología e Historia.

Carrasco, P., and J. Broda (eds.). 1982. *La estratificación social en la Mesoamérica prehispánica*. México, D.F., Instituto Nacional de Antropología e Historia.

Carter, W. 1969. *Old Lands and New Traditions: Kekchi Cultivators in teh Guatemala Lowlands*. Gainesville, University of Florida Press.

Caskie, P. 2000. Back to Basics: Household Food Production in Russia. *Journal of Agricultural Economicz* 51: 196-209.

Castanzo, R. A. 2002. *The Development of Social Complexity in the Formative Period Central Puebla-Tlaxcala Basin, Mexico*. Unpublished PhD thesis, The Pennsylvania State University.

Castanzo, R. A., and J. H. Anderson. 2003. Formative Period lime kilns in Puebla, Mexico. *Mexicon* XXVI: 86-89.

Castanzo, R. A., and J. J. Sheehy. 2004. The Formative Period civic-ceremonial centre of Xochiltenango in Mexico. *Antiquity* 2009.

Castillo, V. M. 1972. Unidades nahuas de medida. *Estudios de Cultura Náhuatl* 10: 195-224.

Clark, C., and M. Haswell. 1967. *The Economics of Subsistence Agriculture*. London, St. Martin's Press.

Cleave, J. H. 1974. *African Farmers: Labor Use in the Development of Smallholder Agriculture*. New York, Praeger Publishers.

Cleveland, M. K., D. W. Stahle, M. D. Therrell, J. Villanueva Diaz, and B. T. Burns. 2003. Tree Ring reconstructed winter precipitation and tropical teleconnections in Durango, Mexico. *Climatic Change* 59: 369-388.

Cline, S. L. 1986. *Colonial Culhuacan, 1580-1600: A Social History of an Aztec Town*. Albuquerque, University of New Mexico Press.

Conde, C., R. Ferrer, and S. Orozco. 2006. Climate and Climate Variability Impacts on Rainfed Agricultural Activities and Possible Adaptation Measures. A Mexican Case Study. *Atmósfera* 19: 181-194.

Conklin, H. C. 1954. An Ethnoecological Approach to Shifting Agriculture. *Transactions of the New York Academy of Sciences* 17: 133-142.

Cook, S. F. 1996. La erosión del suelo y la población del centro de México. In S. F. Cook and W. Borah (eds.), *El pasado de México: Aspectos Sociodemográficos*: 88-171. México, D.F., Fondo de Cultura Económica.

Cortés, H. 1992. *Cartas de Relación*. México, D.F., Editores Mexicanos Unidos, S.A.

Cortés, M. 1865. Carta de don Martín Cortés, segundo marqués del Valle, al rey don Felipe II, sobre los repartimientos y clases de tierras de la Nueva España. In J. Pacheco, D. d. Cárdenas, and L. Torrez de Mendoza (eds.), *Colección de documentos inéditos relativos al descubrimiento, conquista y organización de las antiguas posesiones españolas de América y Oceanía, sacados de los archivos del reino y muy especialmente del de las Indias*, vol. IV. Madrid, Archivo General de Indias.

Cowgill, U. 1962. An agricultural study of the southern Maya Lowlands. *American Anthropologist* 64: 273-286.

Crosson, P. R. 1970. *Agricultural Development and Productivity: Lessons from the Chilean Experience*. Baltimore, The Johns Hopkins Press.

Chance, J. K. 2000. The Noble House in Colonial Puebla, Mexico: Descent, Inheritance, and the Nahua Tradition. *American Anthropologist, New Series* 102: 485-502.

Charlton, T. H. 1970. Contemporary Agriculture of the Valley. In W. T. Sanders, A. Kovar, T. H. Charlton, and R. A. Diehl (eds.), *The Natural Environment, Contemporary Occupation and 16th Century Population of the Valley: The Teotihuacan Valley Project, Final Report*: 253-384. University Park, Ocassional Papers in Anthropology No. 3, The Pennsylvania State University.

Chavero, A. 1979. *El Lienzo de Tlaxcala*. México, D.F., Editorial Innovación, S.A.

Chayanov, A. V. 1966. *The Theory of Peasant Economy*. Homewood, The American Economic Association.

Chibnik, M. 1984. A Cross-cultural Examination of Chayanov's Theory. *Current Anthropology* 25: 335-340.

—. 1987. The Economic Effects of Household Demography: A Cross-Culture Assessment of Chayanov's Theory. In M. D. Maclachlan (ed.) *Household Economies and Their Transformations*: 74-106. Lanham, Univeristy Press of America.

Chimalpáhin, D. 1998. *Las ocho relaciones y el memorial de Colhuacan*. Vol. I-II. México, D.F., Consejo Nacional para la Cultura y las Artes.

Chisholm, M. 1970. *Rural Settlement and Land Use: An Essay in Location*. Chicago, Aldine Publishing Company.

D'Altroy, T. N., and T. K. Earle. 1985. Staple Finance, Wealth Finance, and Storage in the Inka Political Economy. *Current Anthropology* 26: 187-206.

Dahlin, B. H., T. Beach, S. Luzzadder-Beach, D. Hixson, S. Hutson, A. Magnoni, E. Mansell, and D. E. Mazeau. 2005. Reconstructing agricultural Self-Sufficiency at Chunchucmil, Yucatan, Mexico. *Ancient Mesoamerica* 16: 229-247.

de Leon, N., and J. G. Coors. 2002. Twenty-Four Cycles of Mass Selection for Prolificacy in the Golden Glow Maize Population. *Crop Science* 42: 325-333.

de Leon, N., J. G. Coors, and S. M. Kaeppler. 2005. Genetic Control of Prolificacy and Related Traits in the Golden Glow Maize Population: II. Genotypic Analysis. *Crop Science* 45: 1370-1378.

De Vries, E. 1952. *The Agricultural Economy of Chile*. Report of a Mission Organized by the International Bank for reconstruction and Development and the Food and Agriculture Organization of the United Nations.

Dean, J. S., D. M. Meko, and T. W. Swetnam. 1996. *Tree rings, environment, and humanity: proceedings of the international conference, Tucson, Arizona, 17-21 May 1994*. Tucson, University of Arizona.

Dincauze, D. F. 2000. *Environmental Archaeology: principles and practice*. Cambridge, Cambridge University Press.

Dopico, E., A. R. Linde, and E. García-Vazquez. 2009. Traditional and Modern Practices of Soil Fertilization: Effects on Cadmium Pollution of River Ecosystems in Spain. *Human Ecology* 37: 235-240.

Douglas, M. W., R. A. Maddox, and K. Howard. 1993. The Mexican Monsoon. *Journal of Climate* 6: 1665-1677.

Dove, M. R. 1983. Ethnographic Atlas of Ifugao: Implications for Theories of Agricultural Evolution in Southeast Asia. *Current Anthropology* 24: 516-519.

Drucker, P., and R. F. Heizer. 1960. A Study of Milpa System of La Venta and Its Archaeological Implications. *Southwestern Journal of Anthropology* 16: 36-45.

Dunning, N. P. 1996. A Reexamination of Regional Variability in the Pre-Hispanic Agricultural Landscape. In S. L. Fedick (ed.) *The Managed Mosaic: Ancient Maya Agriculture and Resource Use*: 53-68. Salt Lake City, University of Utah Press.

Durán, D. 2006 [1579]. *Historia de las Indias de Nueva España e Islas de la Tierra Firme*. México, D.F., Editorial Porrúa, S.A. de C.V.

Durrenberger, E. P. 1980. Chayanov's Economic Analysis in Anthropology. *Journal of Anthropological Research* 36: 133-148.

Dyckerhoff, U. 1978. La Época Prehispánica. In H. J. Prem (ed.) *Milpa y hacienda: tenencia de la tierra indígena y española en la cuenca del Alto Atoyac, Puebla, México (1520-1650)*: 18-34. Wiesbaden, Steiner.

Eakin, H. 2000. Smallholder Maize Production and Climatic Risk: A Case Study From Mexico. *Climatic Change* 45: 19-36.

Earle, T. K. 2000. Archaeology, Property, and Prehistory. *Annual Review of Anthropology* 29: 39-60.

Erffa, A. v., W. Hilger, K. Knoblich, and R. Weyl. 1977. *Geologie des Hochbeckens von Puebla-Tlaxcala und seiner Umgebung*. Wiesbaden, Steiner.

Evans, S. T. 1990. The Productivity of Maguey Terrace Agriculture in Central Mexico During the Aztec Period. *Latin American Antiquity* 1: 117-132.

—. 2001. Aztec-Period Political Organization in the Teotihuacan Valley: Otumba as a city-state. *Ancient Mesoamerica* 12: 89-100.

FAO. 2001. *Grape Production in the Asia-Pacific Region*. Bangkok, Food and Agriculture Organization of the United Nations Regional Office for Asia and the Pacific.

Fargher, L. F. 2007. In the Shadow of Popocatepetl: Archaeological Survey and Mapping at Tlaxcala, México. *FAMSI*.

Fargher, L. F., R. E. Blanton, and V. Y. Heredia. 2010a. Egalitarian Ideology and Political Power in Prehispanic

Central Mexico: The Case of Tlaxcallan. *Latin American Antiquity* 21: 227-251.

Fargher, L. F., R. E. Blanton, V. Y. Heredia, J. Millhauser, N. Xiutecuhtli, and L. Overholtzer. 2010b. Tlaxcallan: the archaeology of an ancient republic in the New World. *Antiquity* 84: 1-15.

Fargher, L. F., V. Y. Heredia, and R. E. Blanton. 2011. Alternative pathways to power in late Postclassic Highland Mesoamerica. *Journal of Anthropological Archaeology* 30: 306-326.

Fedick, S. L. 1996. An Interpretative Kaleidoscope: Alternative Perspectives on Ancient Agricultural Landscapes of the Maya Lowlands. In S. L. Fedick (ed.) *The Managed Mosaic: Ancient Maya Agriculture and Resource Use*: 107-131. Salt Lake City, University of Utah Press.

Feinman, G. M., and L. M. Nicholas. 2004. Unraveling he Prehispanic Highland Mesoamerican Economy: Production, Exchange, and Consumption in the Classic Period Valley of Oaxaca. In G. M. Feinman and L. M. Nicholas (eds.), *Archaeological Perspectives on Political Economies*: 167-188. Salt Lake City, The University of Utah Press.

Feinman, G. M., L. M. Nicholas, and H. R. Haines. 2007. Classic Period Agricultural Intensification and Domestic Life at El Palmillo, Valley of Oaxaca, Mexico. In T. L. Thurston and C. T. Fisher (eds.), *Seeking a Richer Harvest: The Archaeology of Subsistence Intensification, Innovation, and Change*: 23-61. Verlag, Springer.

Fernández, A. S. 1977. *Bosquejo geológico de la Sierra de Tecamachalco en el Estado de Puebla*. Unpublished Licenciatura thesis, Instituto Politécnico Nacional.

Fernández de Oviedo, G. 1959. *Historia General y Natural de la Indias: Vol. IV*. Vol. IV. Madrid, Gráficas ORBE.

Fernández, L. M., and R. F. Wasserstrom. 1977. Los Municipios Alteños de Chiapas (México) y sus Relaciones con la Economía Regional: Dos Estudios de Caso. *Estudios Sociales Centroamericanos*: 29-69.

Flores, L., G. Jiménez, R. G. Martínez, R. E. Chávez, and D. Silva. 2006. Study of geothermal water intrusion due to groundwater exploitation in the Puebla Valley aquifer system, México. *Hydrogeology Journal* 14: 1216-1230.

Florescano, E. 1980. *Análisis Histórico de las Sequías en México*. México, D.F., Secretaría de Agricultura y Recursos Hidráulicos.

—. 2006. El Altepetl. *Fractal* 42.

Fowler, M. L. 1987. Early Water Management at Amalucan, State of Puebla, Mexico. *National Geographic Research* 3: 52-68.

Freeman, J. D. 1955. *Iban Agriculture: A Report on the Shifting Cultivation of Hill Rice by the Iban Sarawak*. Vol. No. 18. London, Her Majesty's Stationery Office.

Fuentes, L. 1972. *Regiones naturales del Estado de Puebla*. México, Universidad Nacional Autónoma de México, Instituto de Geografía.

Gámez, A. 2003. *Los popolocas de Tecamachalco-Quecholac*. Puebla, Benemérita Universidad Autónoma de Puebla.

Garavaglia, J. C., and J. C. Grosso. 1990. Mexican Elites of a Provincial Town: The Landowners of Tepeaca (1700-1870). *The Hispanic American Historical Review* 70: 255-293.

García, V., J. M. Pérez, and A. Molina del Villar. 2003. *Desastres agrícolas en México: catálogo histórico. Tomo 1. Épocas Prehispánica y Colonial (958-1822)*. México, D.F., Centro de Investigaciones y Estudios Superiores en Antropología Social.

García Cook, Á. 1981. The Historical Importance of Tlaxcala in the Cultural Development of the Central Highlands. In J. A. Sabloff (ed.) *Archaeology* 244-276. Austin, University of Texas Press.

—. 1985. Historia de la Tecnología Agrícola en el Altiplano central desde el Principio de la agricultura hasta el Siglo XIII. In T. Rojas and W. T. Sanders (eds.), *Historia de la Agricultura: Época Prehispánica Siglo XVI*, vol. II: 7-75. México, D.F., Instituto Nacional de Antropología e Historia.

García Cook, Á., and B. L. Merino. 1986. Integración y Consolidación de los Señoríos Tlaxcala; Siglos IX a XVI. In Gobierno del Estado de Tlaxcala (ed.), *Historia y Sociedad en Tlaxcala: Memorias del Primer Simposio Internacional de Investigaciones Socio-Históricas sobre Tlaxcala*: 23-29. México, D.F., Gobierno del Estado de Tlaxcala.

García, B. 1998. El *altépetl* o *pueblo de indios*: Expresión básica del cuerpo político mesoamericano. *Arqueología Mexicana* 32: 58-65.

Garraty, C. P. 2006. Aztec Teotihuacan: Political Processes at a Postclassic and Early Colonial City-State in the Basin of Mexico. *Latin American Antiquity* 17: 363-387.

Gaspar, E. Á. 2010. *Caracterización, rendimiento de maízces nativos y descripción de las unidades de producción en el municipio de Molcaxac, Pue.* . Puebla, Unpublished Master Thesis, Colegio de Postgraduados.

Geertz, C. 1963. *Agricultural Involution: The Processes of Ecological Change in indonesia*. Berkeley, University of California Press.

Gibson, C. 2003. *Los Aztecas Bajo el Dominio Español: 1519-1810*. México, D.F., Siglo Veintiuno Editores S.A. de C.V.

Gil, A., P. A. López, A. Muñoz, and H. López. 2004. Variedades criollas de maíz (Zea mays L.) en el estado de Puebla, México: diversidad y utilización. In J. L. Chávez, J. Tuxill, and D. I. Jarvis (eds.), *Manejo de la diversidad de los cultivos en los agroecosistemas tradicionales*. Cali, Instituto Internacional de Recursos Fitogenéticos.

Gimenez, C., F. Orgaz, and E. Fereres. 1997. Productivity in Water-Limited Environments: Fryland Agricultural Systems. In L. E. Jackson (ed.) *Ecology in Agriculture*. San Diego, Academic Press.

Gliessman, S. R. 2000. *Agroecology: Ecological Processes in Sustainable Agriculture*. Boca Raton, Lewis Publishers.

Glockner, J. 1991. Los Volcanes en los Sueños del "tiempero". *Primer Coloquio sobre Puebla*: 32-45.

Puebla, Comisión Puebla V Centenario, Gobierno del Estado de Puebla.

González de Cossío, F. 1952. *El Libro de las tasaciones de Puebla de la Nueva España: Siglo XVI*. México, D.F., Archivo General de la Nación. Editorial E.C.L.A.L.

González, A., L. M. Vázquez, J. Sahagún, J. E. Rodríguez, and D. d. J. Pérez. 2007. Rendimiento del mapiz de temporal y su relación con la pudrición de la mazorca. *Agricultura Técnica en México* 33: 33-42.

Gónzalez, A. 2003. *Cultura y agricultura: transformaciones en el agro mexicano*. México, D.F., Unversidad Iberoamericana.

Goudie, A. 2001. *The Human Impact on the Natrual Environment*. Cambridge, The MIT Press.

Granados, R., T. Reyna, J. Soria, and Y. Fernández. 2004. Aptitud agroclimática en la mesa central de Guanajuato, México. In*vestigaciones Geográficas* agosto: 24-35.

Green, S. W. 1980. Towards a General Model of Agricultural Systems. In M. B. Schiffer (ed.) *Advances in Archaeological Method and Theory*, vol. 3. New York, Academic Press.

Gutierrez, G. 2003. Territorial Structure and Urbanism in Mesoamerica: The Huaxtec and Mixtec-Tlapanec-Nahua Cases. In W. T. Sanders, A. G. Mastache, and R. H. Cobean (eds.), *El Urbanismo en Mesoamérica*: 85-118. México, D.F., Instituto Nacional de Antropología e Historia.

Halstead, P., and J. O'Shea. 1989. Introduction: cultural responses to risk and uncertainty. In P. Halstead and J. O'Shea (eds.), *Bad Year Economics: Cultural Responses to Risk and Uncertainty*: 1-7. Cambridge, Cambridge University Press.

Hammel, E. A. 2005. Chayanov revisited: A model for the economics of complex kin units. *Proceedings of the National Academy of Science* 102: 7043-7046.

Hassig, R. 1981. The Famine of One Rabbit: Ecological Causes and Social Consequences of Pre-Columbian Calamity. *Journal of Archaeological Research* 37: 172-182.

—. 1982. Periodic Markets in Precolumbian Mexico. *American Antiquity* 47: 346-355.

Hayden, B. 2001. Richman, Poorman, Beggarman, Chief: The Dynamics of Social Inequality. In G. Feinman and D. Price (eds.), *Archaeology at the Millennium: A Sourcebook*: 231-272. New York, Kluwer Academic Press.

Hernandez Xolocotzi, E. 1965. *Graneros para maíz en México a través de los siglos*. Chapingo, Colegio de Posgraduados de la Escuela Nacional de Agricultura.

Hicks, F. 1982. Tetzcoco in the Early 16th Century: The State, the City, and the "Calpolli". *American Ethnologist* 9: 230-249.

—. 2009. Land and Succession in the Indigenous Noble Houses of Sixteenth-Century Tlaxcala. *Ethnohistory* 56: 569-588.

Hirth, K. G. 1996. Political Economy and Archaeology: Perspectives on Exchange and Production. *Journal of Archaeological Research* 4: 203-239.

—. 1998. The Distributional Approach: A New Way to Identify Marketplace Exchange in the Archaeological Record. *Current Anthropology* 39: 451-476.

—. 2003. The Altepetl and Urban Structure in Prehispanic Mesoamerica. In W. T. Sanders, A. G. Mastache, and R. H. Cobean (eds.), *El Urbanismo en Mesoamérica*, vol. 1: 57-84. Mexico, D.F., Instituto Nacional de Antropología e Historia and The Pennsylvania State University.

—. 2007. *Housework: Craft Production, Risk, and Domestic Economy in Mesoamerica*. 2007 meeting of the Society for American Archaeology, Austin, Texas.

—. 2009. Craft Production, Household Diversification, and Domestic Economy in Prehispanic Mesoamerica. *Archaeological Papers of the American Anthropological Association* 19: 13-32.

— 2010. "The Merchant's World: Commercial Diversity and the Economics of Interregional Exchange in Highland Mesoamerica." Dumbarton Oaks.

Hoque, M. Z. 1984. *Cropping Systems in Asia: On Farm Research and Management*. Manila, International Rice Research Institute.

Hortelano, R., A. Gil, A. Santacruz, S. Miranda, and L. Córdova. 2008. Diversidad morfológica de maíces nativos del Valle de Puebla. *Agricultura Técnica en México* 34: 189-200.

Hughes, M. K., P. M. Kelly, and J. R. Pilcher. 2009. *Climate from Tree Rings*. Cambridge, Cambridge University Press.

Iglesias, M. 2000. Tierra y estratificación social indígena en Cuauhtinchan. (Siglo XVI). *Cuadernos de la Facultad de Humanidades y Ciencias Sociales* noviembre: 251-281.

INEGI. 2000. *Síntesis geográfica del estado de Puebla: libro electrónico*. México, D.F., Instituto Nacional de Estadística y Geografía.

Instituto Nacional de Investigaciones Forestales, A., y Pecuarias (INIFAP). 1997. *Guía para la asistencia técnica agrícola en el área de influencia del campo experimental Tecamachalco*. Tecamachalco, Puebla, Secretaría de Agricultura, Ganadería, y Desarrollo Rural y El Instituto Nacional de Investigaciones Forestales, Agrícolas, y Pecuarias.

Jackson, I. J. 1978. Local differences in the Patterns of Variability of Tropical Rainfall: Some Characteristics and Implications. *Journal of Hydrology* 38: 273-278.

Jaén, M. T., C. Serrano, and J. Comas. 1976. Data antropométrica de algunas poblaciones indígenas mexicanas. *Anales de Antropología* 13.

Jalpa, T. 2009. *La sociedad indígena en la región de Chalco durante los siglos XVI y XVII*. México, D.F., Instituto Nacional de Antropología e Historia.

Janusek, J. W., and A. L. Kolata. 2004. Top-down or bottom-up: rural settlement and raised field agriculture in the Lake Titicaca Basin, Bolivia. *Journal of Anthropological Archaeology* 23: 404-430.

Jiménez, W. 1995. El enigma de los olmecas. In Á. García Cook, B. L. Merino, and L. Mirambell (eds.), *Antología de Cacaxtla*, vol. I: 73-109. México, D.F., Instituto Nacioalto de Antropología e Historia.

Johnson, A. W., and T. Earle. 1987. *The Evolution of Human Societies: From Foraging Group to Agrarian State*. Stanford, Stanford University Press.

Jordán, A., L. M. Zavala, A. L. Nava, and N. Alanís. 2009. Occurence and hydrological effects of water repellancy in different soil and land use types in Mexican volcanic highlands. *Catena* 79: 60-71.

Joyce, A. A. 2001. Commoner Power: A Case Study from the Classic Period Collapse on the Oaxaca Coast. *Journal of Archaeological Method and Theory* 8: 343-385.

Kalton, G. 1983. In*troduction to Survey Sampling*. Vol. 35. Newbury Park, Sage Univeristy Paper.

Kass, D. C. L., and E. Somarriba. 1999. Traditional fallows in Latin America. *Agroforestry Systems* 47: 13-36.

Kay, G. 1964. "Aspects of Ushi Settlement History," in *Geographers and the Tropics*. Edited by R. W. Steele and R. M. Prothero, pp. 235-260. London: Longman.

Killion, T. W. 1990. Cultivation Intensity and Residential Site Structure: An Ethnoarchaeological Examination of Peasant Agriculture in the Sierra de los Tuxtlas, Veracruz, Mexico. *Latin American Antiquity* 1: 191-215.

Kirch, P. V. 1994. *The Wet and the Dry: Irrigation and Agricultural Intensification in Polynesia*. Chicago, The University of Chicago Press.

Kirchhoff, P., L. Odena, and L. Reyes. 1976. *Historia Tolteca Chichimeca*. México, D.F, INAH-SEP.

Kirkby, A. V. T. 1973. *The Use of Land and Water Resources in the Past and Present Valley of Oaxaca, Mexico*. Vol. 1. Ann Arbor, Memoirs of the Museum of Anthropology no. 5, University of Michigan.

Klink, H. K. 1973. La división de la vegetación natural en la región de Puebla-Tlaxcala. *comunicaciones* 7: 25-30.

Kowal, J. M., and A. H. Kassam. 1978. *Agricultural Ecology of Savanna: A Study of West Africa*. Oxford, Clarendon Press.

Kowalewski, S. A., and R. D. Drennan. 1989. *Prehispanic settlement patterns in Tlacolula, Etla, and Ocotlan, the Valley of Oaxaca, Mexico*. Ann Arbor, Regents of the University of Michigan, the Museum of Anthropology.

Kuchko, A. A. 1998. Current situation and prospects of potato production in Ukraine. *Bulletin OEPP/EPPO* 28: 433-437.

Lagunas, Z., and S. López. 2004. Antropología Física en grupos humanos de filiación Otopame. *Ciencia Ergo Sum* 11: 47-58.

Lailand, K. N., and G. R. Brown. 2006. Niche Construction, Human Behavior, and the Adaptive-Lag Hypothesis. *Evolutionary Anthropology* 15: 95-104.

Lamb, J. F. S., C. C. Sheaffer, and D. A. Samac. 2003. Population Density and Harvest Maturity Effects on Leaf and Stem Yield in Alfalfa. *Agronomy Journal* 95: 635-641.

Lane, M., R. Aguirre, and J. González. 1997. Producción campesina del maíz en San Lorenzo Tenochtitlán. In A. Cyphers (ed.) *Población, subsistencia y medioambiente en San Lorenzo Tenochtitlán*: 55-74. México, D.F., Universidad Nacional Autónoma de México.

Lara, E., M. Aliphat, and B. Ramírez. 2002. *Zentli: La agricultura del maíz en una comunidad nahua de La Malinche, Tlaxcala*. Tlaxcala, CONACULTA PACMYC.

Lauer, W. 1979. Medioambiente y desarrollo cultural en la región de Puebla-Tlaxcala. *comunicaciones* 16: 29-54.

Leander, B. 1967. *Códice de Otlazpan*. México D.F., Instituto Nacional de Antropología e Historia.

Leyden, B. W., M. Brenner, T. Whitmore, J. H. Curtis, D. R. Piperno, and B. H. Dahlin. 1996. A Record of Long- and-Short-Term Climatic Variation from Northwest Yucatán: Cenote and San José Chulchacá. In S. L. Fedick (ed.) *The Managed Mosaic: Ancient Maya Agriculture and Resource Use*. Salt Lake city, University of Utah Press.

Liverman, D. M., and K. L. O'Brian. 1991. Global warming and climatic change in Mexico. *Global Environmental Change*: 351-364.

Lockhart, J. 1992. *The Nahuas After the Conquest: A Social and Cultural History of the Indians of Central Mexico, Sixteenth Through Eighteenth Centuries*. Stanford, Stanford University Press.

Logan, M. H., and W. T. Sanders. 1976. The Model. In E. R. Wolf (ed.) *The Valley of Mexico: Studies in Pre-Hispanic Ecology and Society*: 31-58. Albuquerque, University of New Mexico Press.

Lohse, J. C., and N. Gonlin. 2007. Preface. In N. Gonlin and J. C. Lohse (eds.), *Commoner Ritual and Ideology in Ancient Mesoamerica*: xvii-xxxix. Boulder, University Press of Colorado.

Loker, W. 1989. Contemporary land use and prehistoric settlement: an ethnoarchaeological approach. In K. G. Hirth, G. Lara, and G. Hasemann (eds.), *Archaeological Research in the El Cajón Region, Volume I: Prehistoric Cultural Ecology*: 135-186. Pittsburgh, University of Pittsburgh Press.

López, A. 2000. *Dos Mil Años de Tradición Agrícola: Tecnología y Organización Social durante el Formativo Terminal en Tetimpa, Puebla*. Unpublished Licenciatura thesis, Universidad de las Américas, Puebla.

—. 2012. Conquistas y macehuales: la fuente de riqueza en Cuauhtinchan y Tepeaca durante la época prehispánica. *Revista Teccalli* 3: 21-27.

López, A., and K. G. Hirth. 2012. Terrazguero Smallholders and the Function of Agricultural Tribute in Sixteenth-Century Tepeaca, Mexico. *Mexican Studies / Estudios Mexicanos* 28: 73-93.

López Austin, A. 1984. *Cuerpo humano e ideología: las concepciones de los antiguos nahuas*. Vol. I y II. México, D.F., Universidad Nacional Autónoma de México.

Luna, M., and R. Gaytán. 2001. Rendimiento de maíz de temporal con tecnología tradicional y recomendada. *Agricultura Técnica en México* 27.

Mabry, J. B., and D. A. Cleveland. 1996. The relevance of Indeigenous Irrigation: A Comparative Analysis of Sustainability. In J. B. Mabry (ed.) *Canals and Communities: Small-Scale Irrigation Systems*: 227-260. Tucson, University of Arizona Press.

Magaña, V., J. A. Amador, and S. Medina. 1999. The midsummer drought over Mexico and Central America. *Journal of Climate* 12: 1577-1588.

Magaña, V., J. L. Pérez, J. L. Vázquez, E. Carrisoza, and J. Pérez. 2004. El Niño y el clima. In V. Magaña Rueda (ed.) *Los impactos del niño en México*: 23-68. México, D.F., Centro de Ciencias de la Atmósfera, Universidad Nacional Autónoma de México, Secretaría de Gobernación.

Maiti, R., and P. Wesche-Ebeling. 1998. *Maize Science*. Enfield, Science Publishers Inc.

Maldonado, B. 1997. *Las figurillas formativas del área de Acatzingo-Tepeaca*. Unpublished Lecenciatura thesis, Unversidad de las Américas, Puebla.

Marcus, J., and G. M. Feinman. 1998. Introduction. In G. M. Feinman and J. Marcus (eds.), *Archaic States*: 3-13. Sante Fe, School of American Research Press.

Maro, P. S. 1996. I*ntegrated Land Use Planning in Semi-Arid Areas: report on the Second SADC-ELMS Practical Workshop*. SADC Environment and Land Management Sector.

Martínez, H. 1984a. *Colección de documentos coloniales de Tepeaca*. México, D.F., Instituto Nacional de Antropología e Historia.

—. 1984b. *Tepeaca en el Siglo XVI: tenencia de la tierra y organización de un señorío*. México, D.F., Ediciones de la Casa Chata, Centro de Investigaciones y Estudios Superiores en Antropología Social.

—. 1994a. *Codiciaban la tierra: el despojo agrario de los señoríos de Tecamachalco y Quecholac (Puebla, 1520-1650)*. México, D.F., Centro de Investigaciones y Estudios Superiores en Antropología Social.

—. 1994b. La Conquista de Tepeyacac: una Estrategia Política de Expansión del Imperio Mexica. *Revista Mexicana de Estudios Antropológicos* XL: 133-168.

—. 2001. Calpulli ¿Otra acepción de *teccalli*? in A. Escobar Ohmstede and T. Rojas (eds.), *Estructuras y formas agrarias en México: del pasado y del presente*: 25-44. México, D.F., Centro de Investigaciones y Estudios Superiores en Antropología Social.

Matías, M. 1984. *Medidas indígenas de longitud (en documentos de la Ciudad de México del siglo XVI)*. México, D.F., CIESAS.

McCaa, R. 1995. Spanish and Nahuatl Views on Smallpox and Demographic Catastrophe in Mexico. *Journal of Interdisciplinary History* 25: 397-431.

McGregor, G. R., and S. Nieuwolt. 1998. *Tropical Climatology: an Introduction to the Climates of the Low Latitudes*. New York, John Wiley & Sons.

Medina, M. 2001. *Las Cuevas de Acatzingo-Tepeaca, Puebla: estudio arqueológico, etnohistórico y etnográfico*. Unpublished Licenciatura thesis, Escuela Nacional de Antropología e Historia.

Merino, B. L., and Á. García Cook. 1998. Los señoríos prehispánicos de la provincia de Tlaxcala según la arqueología. In Gobierno del Estado de Tlaxcal (ed.), *Coloquio sobre la historia de Tlaxcala*: 87-104. Tlaxcala, Gobierno Constitucional del Estado de Tlaxcala.

Minnis, P. 1985. *Social Adaptation to Food Stress: A Prehistoric Southwestern Example*. Chicago, University of Chicago Press.

Molina, F. A. d. 2008 [1571]. *Vocabulario en Lengua Castellana y Mexicana y Mexicana y Castellana*. México, D.F., Editorial Porrúa.

Molina, F. d. 1985 [1580]. Relación de Tepeaca y su partido. In R. Acuña (ed.) *Relaciones Geográficas del siglo XVI: Tlaxcala*, vol. 2: 217-260. México, D.F., Universidad Nacional Autónoma de México.

Moll, R. H., and W. D. Hanson. 1984. Comparisons of Effects of Intrapopulation vs. Interpopulation Selection in Maize. *Crop Science* 24: 1047-1052.

Monteforte, M. 1959. *Guatemala. Monografía Sociológica*. México, D.F., Universidad Nacional Autónoma de México.

Morley, S. G., and G. W. Brainerd. 1968. *The Ancient Maya*. Standford, Standford University Press.

Mountjoy, J. B., and D. L. Peterson. 1973. *Man and land at prehistoric Cholula*. Nashville, Vanderbilt University.

Moyes, H., J. J. Awe, G. A. Brook, and J. W. Webster. 2009. The Ancient Maya Cult: Late Classic Cave Use in Belize. *Latin American Antiquity* 20: 175-206.

Munson-Scullin, W., and M. Scullin. 2005. Potential Productivity of Midwestern Native American Gardens. *Plains Anthropologist* 50: 9-21.

Muñoz Camargo, D. 1998 [1580]. *Historia de Tlaxcala (Ms. 210 de la Biblioteca Nacional de París)*. Tlaxcala, Universidad Autónoma de Tlaxcala.

Murtha Jr., T. M. 2002. *Land and Labor: Classic Maya Terraced Agriculture at Caracol, Belize*. Unpublished PhD, The Pennsylvania State University.

Mwalukisa, P. 2008. Changing attitudes to night-soil in Tanzania. *Laeisa Magazine* 242: 26-27.

Nations, J. D., and R. B. Nigh. 1980. The Evolutionary Potential of Lacandon Maya Sustained-Yield Tropical Forest Agriculture. *Journal of Anthropological Research* 36: 1-30.

ne Nsaku, N., and G. C. W. Ames. 1982. *Constraints to Maize Production in Zaire*. Athens, The University of Georgia, College of Agriculture Experiment Stations.

Netting, R. M. 1993. *Smallholders, Householders: Farm Families and the Ecology of Intensive, Sustainable Agriculture*. Stanford, Stanford University Press.

Ng'ong'ola, D. H., R. N. Kachule, and P. H. Kabambe. 1997. *The Maize Market in Malawi*. Lilongwe, Malawi, Agricultural Policy Research Unit, Bunda College of Agriculture.

Nicholas, L. M. 1989. Land use in Prehispanic Oaxaca. In S. A. Kowalewski, G. M. Feinman, L. Finsten, R. E. Blanton, and L. M. Nicholas (eds.), *Monte Alban's Hinterland, Part II: Prehispanic Settlement Patterns in Tlacolula, Etla and Ocotlan, The Valley of Oaxaca, Mexico*: 449-504. Ann Arbor, University of Michigan.

Nichols, D. L. 1987. Risk and Agricultural Intensification during the Formative Period in the Northern Basin of Mexico. *American Anthropologist, New Series* 89: 596-616.

Nichols, G. 2009. *Sedimentology and Stratigraphy*. Oxford, Wiley-Blackwell.

Niembro, A. 1986. Árboles y Arbustos útiles de México: naturales e introducidos. México, D.F., Editorial LIMUSA.

Nigh, R. B. 1976. *Evolutionary Ecology of Maya Agriculture in Highland Chiapas, Mexico*. Unpublished Ph.D. dissertation, Department of Anthropology. Stanford, Stanford University.

Norman, M. J. T. 1979. *Annual cropping systems in the tropics : an introduction*. Gainesville, University Presses of Florida.

O'Sullivan, D., and D. Unwin. 2003. *Geographic Information Analysis*. Hoboken, John Wiley & Sons, Inc.

Ohngemach, D., and H. Straka. 1978. La historia de la vegetación en la región de Puebla-Tlaxcala durante el Cuaternario Tardío. *comunicaciones* 15: 189-204.

Olivera, M. 1973. La Estructura social de Tecali en el Siglo XVI. *comunicaciones* 8: 31-35.

—. 1978. *Pillis y macehuales, las formaciones sociales y los modos de producción de Tecali del Siglo XII al XVI*. México, D.F., Editorial de la Casa Chata.

Padoch, C., e. Harwell, and A. Susanto. 1998. Swidden, Sawah, and In-Between: Agricultural Transformation in Borneo. *Human Ecology* 26: 3-20.

Palerm, Á. 1955. The Agricultural Basis of Urban Civilization in Mesoamerica. In J. H. Steward (ed.) *Irrigation Civilizations: A Comparative Study*. Washington, D.C., Pan American Union.

—. 1967. Agricultural Systems and food Patterns. *Handbook of Middle American Indians*, vol. 6: 26-52. Austin, University of Texas Press.

—. 1972. *Agricultura y sociedad en Mesoamérica*. México, D.F., Sepsetentas 55.

—. 1990. Aspectos agrícolas del desarrollo de la civilización prehispánica en Mesoamérica (evidencias entnográficas). In C. Viqueira (ed.) *México prehispánico: Ensayos sobre evolucipon y ecología*. México, D.F., Consejo Nacional para la Cultura y las Artes.

Parsons, J. R. 1991. Political Implications of Prehispanic Chinampa Agriculture in the Valley of Mexico. In H. R. Harvey (ed.) *Land and Politics in the Valley of Mexico: A Two-Thousand-Year Perspective*: 17-43. Albuquerque, University of New Mexico Press.

—. 1992. El Papel de la Agricultura Chinampera en el Abasto Alimenticio de Tenochtitlán. In C. J. González (ed.) *Chinampas Prehispánicas*: 207-244. México, D.F., INAH.

Parsons, J. R., and M. H. Parsons. 1990. *Maguey Utilization in Highland Central Mexico: An Archaeological Ethnography*. Ann Arbor, University of Michigan.

Pataky, J. K., and M. A. Chandler. 2003. Production of huitlacoche, *Ustilago maydis*: timing inoculation and controlling pollination. *Mycologia* 95: 1261-1270.

Patrick, L. 1977. *Cultural Geography of the Use of Seasonally Dry, Sloping Terrain: The Metepantli Crop Terraces of Central Mexico*, University of Michigan.

Peralta, A. R., V. O. Magaña, A. D. Matthias, and J. d. J. Luna. 2008. Temporal and spatial behavior of temperature and precipitation during the canícula (midsummer drought) under El Niño conditions in central México. *Atmósfera* 21: 265-280.

Perkins, S. M. 2007. The House of Guzmán: An Indigenous Cacicazgo in Early Colonial Central Mexico. *Culture and Agriculture* 29: 25-42.

Pétrequin, P. (ed.) 1994. *8000 Años de la Cuenca de Zacapu: evolución de los paisajes y primeros desmontes*. México, D.F., Centre D´Études Mexicaines et Centraméricaines México.

Petrovici, D. A., and M. Gorton. 2005. An Evaluation of the Importance of subsistence food production for assessments of poverty and policy targeting: Evidence from Romania. *Food Policy* 30: 205-223.

Piesse, J., H. S. von Bach, C. Thirtle, and J. Van Zyl. 2000. Farm Size and Efficiency of Smallholder Agriculture. In C. Thirtle, J. van Zyl, and N. Vink (eds.), *South African Agriculture at the Crossroads: An Empirical Analysis of Efficiency, Technology and Productivity*, vol. 133-148. New York, St. Martin's Press, LLC.

Pimentel, D., and W. Dazhong. 1990. Technological Changes in Energy Use in U.S. Agricultural Production. In C. R. Carroll, J. H. Vandermeer, and P. M. Rosset (eds.), *Agroecology*: 147-164. New York, McGraw-Hill Publishing Company.

Plunket, P. 1990. Arqueología y etnohistoria en el Valle de Atlixco. *Notas Mesoamericanas* 12: 3-18.

Plunket, P., and G. Uruñuela. 1994. The impact of the Xochiyaoyotl in southwestern Puebla. In M. G. Hodge and M. E. Smith (eds.), *Economies and Polities in the Aztec Realm, Studies on Culture and Society 6*: 433-446. Albany, Institute for Mesoamerican Studies, State University of New York at Albany.

—. 1998. Preclassic Household Patterns Preserved under Volcanic Ash at Tetimpa, Puebla, Mexico. *Latin American Antiquity* 9: 287-309.

—. 2005. Recent Research in Puebla Prehistory. *Journal of Archaeological Research* V13: 89-127.

Pomar, J. B. 1941. Relación de Tezcoco. *Relaciones de Texcoco y de la Nueva España*. México, D.F., Editorial Salvador Chavez Hayhoe.

Prändl-Zika, V. 2008. From subsistence farming towards a multifunctional agriculture: Sustainability in the Chinese rural reality. *Journal of Environmental Management* 87: 236-248.

Preciado, J. 1976. Una Colonia Tzeltal en la Selva Lacandona Chiapaneca: Aspectos Socio-Economicos de su Relación con el Ecosistema. In E. Hernández (ed.) *Agroecosistemas de México: contribución a la enseñanza, la investigación y la divulgación agrícola*. Chapingo, Universidad Autónoma de Chapingo.

Prem, H. J. 1978. *Milpa y hacienda: tenencia de la tierra indígena y española en la cuenca del Alto Atoyac, Puebla, México (1520-1650)*. Wiesbaden, Steiner.

Prior, C. L., and W. A. Russell. 1975. Yield Performance of Nonprolific and Prolific Maize Hybrids at Six Plant Densities. *Crop Science* 15: 482-486.

Puleston, D. E. 1982. Appendix 2: The Role of Ramon in Maya Subsistence. In K. V. Flannery (ed.) *Maya Subsistence: Studies in Memory of Dennis E. Puleston*: 353-366. New York, Academic Press.

Putz, M. K., and E. L. Taylor. 1996. Wound Response in Fossil Trees from Antartica and its Potential as a Paleoenvironmental Indicator. *IAWA Journal* 17: 77-88.

Rapp Jr., G., and C. L. Hill. 1998. *The Earth-Science Approach to Archaeological Interpretation*. New Haven, Yale University Press.

Redfield, R., and A. Villa Rojas. 1935. *Chan Kom: A Maya Village*. Washington, D.C., Carnegie Institution.

Redman, C. L. 1999. *Human Impact on Natural Environments*. Tucson, The University of Arizona Press.

Reed, D. M. 1998. *Ancient Maya Diet at Copán, Honduras*. Unpublished PhD thesis, The Pennsylvania State University.

Reina, R. 1967. Milpas and Milperos: implications for prehistoric times. *American Anthropologist* 69: 1-20.

Rennie, S. J. 1991. Subsistence Agriculture Versus Cash Cropping - the Social Repercussions. *Journal of Rural Studies* 7: 5-9.

Restall, M. 2005. *African-Native Relations in Colonial Latin America*. Albuquerque, University of New Mexico Press.

Restall, M., L. Sousa, and K. Terraciano. 2005. *Mesoamerican Voices: native-language writings from Colonial Mexico, Oaxaca, Yucatan, and Guatemala*. Cambridge, Cambridge University Press.

Reyes, L. 1988a. *Cuauhtinchan del siglo XII al XVI: Formación y desarrollo histórico de un señorío prehispánico*. México, D.F., Fondo de Cultura Económica.

—. 1988b. *Documentos sobre tierras y señoríos en Cuauhtinchan*. México, D.F., Fondo de Cultura Económica.

—. 2001. *Documentos históricos Cuahuixmatlac Atetecocho*. Tlaxcala, Departamento de Filosofía y Letras de la Universidad Autónoma de Tlaxcala y el Instituto Tlaxcalteca de la Cultura.

Rijal, K., N. K. Bansal, and P. D. Grover. 1991. Energy and Subsistence Nepalese Agriculture. *Bioresource Technology* 37: 61-69.

Rivera, M. Y., S. Sedov, E. Solleiro, J. Pérez, E. McClung, A. González, and J. Gama Castro. 2007. Degradación ambiental en el valle de Teotihuacan: evidencias geológicas y paleopedológicas. *Boletín de la Sociedad Geológica Mexicana* 59: 203-217.

Robin, C. 2003. New Directions in Classic Maya Household Archaeology. *Journal of Archaeological Research* 11: 307-356.

Rojas, T. 1984. Agricultural Implements in Mesoamerica. In H. R. Harvey and H. J. Prem (eds.), *Explorations in Ethnohistory: Indians of Central Mexico in the Sixteenth Century*: 175-204. Albuquerque, University of New Mexico Press.

—. 1988. *Las Siembras de Ayer: la Agricultura Indígena del Siglo XVI*. México, D.F., CIESAS.

Ruiz, T. 1993. La fenega como unidad de superficie. *Agricultura* 726: 24-28.

Ruthenberg, H. 1971. *Farming Systems in the Tropics*. Oxford, Clarendon Press.

Rzedowski, J. 2006. *Vegetación de México*. México, D.F., Comisión Nacional para el Conocimiento y Uso de la Biodiversidad.

Sahagún, B. d. 1963. *Florentine Codex: General History of the Things of the New Spain*. Santa Fe, The School of American Research and The University of Utah.

Sahlins, M. 1972. *Stone Age Economics*. Chicago, Aldine Publishing Company.

Sanders, W. T. 1957. *Tierra y Agua: A Study of the Ecological Factors in the Development of Mesoamerican Civilizations*, Harvard University.

—. 1976. The Agricultural History of the Basin of Mexico. In E. R. Wolf (ed.) *The Valley of Mexico: studies in Pre-Hispanic Ecology and Society*: 101-159. Albuquerque, University of New Mexico Press.

—. 1992. The Population of the Central Mexican Symbiotic Region, the Basin of Mexico, and the Teotihuacán Valley in the Sixteenth Centrury. In W. M. Denevan (ed.) *The Native Population of the Americas in 1492*: 85-150. Madison, The University of Wisconsin Press.

Sanders, W. T., and T. W. Killion. 1992. Factors Affecting Settlement Agriculture in the Ethnographic and Historic Record of Mesoamerica. In T. W. Killion (ed.) *Gardens of Prehistory: The Archaeology of Settlement Agriculture in Greater Mesoamerica*. Tuscaloosa, University of Alabama Press.

Sanders, W. T., and D. L. Nichols. 1988. Ecological Theory and Cultural Evolution in the Valley of Oaxaca. *Current Anthropology* 29: 33-88.

Sanders, W. T., J. R. Parsons, and R. S. Santley. 1979. *The Basin of Mexico: Ecological Processes in the Evolution of a Civilization*. New York, Academic Press.

Sanders, W. T., and R. S. Santley. 1983. A Tale of Three Cities: Energetics and Urbanization in Pre-Hispanic Central Mexico. In E. Z. Vogt and R. M. Leventhal (eds.), *Prehistoric Settlement Patterns: Essays in Honor of Gordon R. Willey*: 243-291. Albuquerque, University of New Mexico Press.

Sanders, W. T., and D. L. Webster. 1988. The Mesoamerican Urban Tradition. *American Anthropologist, New Series* 90: 521-546.

Schiffer, M. B. 1996. *Formation Processes of the Archaeological Record*. Salt Lake City, University of Utah Press.

Schroeder, S. 1991. Indigenous Sociopolitical Organization in Chimalpin. In H. R. Harvey (ed.) *Land and Politics in the Valley of Mexico: A Two-Thousand-Year Perspective*. Albuquerque, University of New Mexico Press.

Schroeder, S. 1999. Maize Productivity in the Eastern Woodlands and Great Plains of North America. *American Antiquity* 64: 499-516.

Schulze, E.-D., E. Beck, and K. Müller-Hohenstein. 2005. *Plant Ecology*. Heidelberg, Springer.

Seele, E., and K. Tyrakowski. 1985. Cuescomate y zencal en la región Puebla-Tlaxcala, México. *comunicaciones* 5.

Seele, E., K. Tyrakowski, and F. Wolf. 1983. Mercados semanales en la región de Puebla-Tlaxcala. *comunicaciones* Suplemento 9.

Sheehy, J. J., M. Medina, and K. G. Hirth. 1995. I*nforme Técnico sobre la Segunda Temporada del Proyecto Acatzingo-Tepeaca en 1995*. México, D.F., Documento en el Archivo Técnico del INAH.

Sheehy, J. J., M. Medina, B. E. Maldonado, R. A. Constanzo, D. Ebert, A. J. Vonarx, and K. G. Hirth. 1997. I*nforme Técnico sobre la Cuarta Temporada del Proyecto Acatzingo-Tepeaca en 1997*. México, D.F., Documento en el Archivo Técnico del INAH.

Shimada, I. 2007. *Craft production in complex societies: multicraft and producer perspectives*. Salt Lake City, University of Utah Press.

Shuman, M. 1974. *The Town Where Luck Fell: The Economics of Life in a Henequén Zone Pueblo*. New Orleans, Unpublished Ph.D. Dissertation, Anthropology Department, Tulane University.

Shumman, I. 1983. Agricultura y Agricultores en la región de Copán. In C. Baudez (ed.) *Introducción a la Arqueología de Copán*: 195-228. Tegucigalpa, Secretaría de Cultura y Turismo.

Siemens, A. H. 1983. Wetland Agriculture in Pre-Hispanic Mesoamerica. *Geographical Review* 73: 166-181.

Sinclair, T. R., and F. P. Gardner. 1998. Environmental Limits to Plant Production. In T. R. Sinclair and F. P. Gardner (eds.), *Principles of Ecology in Plant Production*: 63-78. New York, CAB International.

Sluyter, A. 1994. Intensive Wetland Agriculture in Mesoamerica: Space, Time, and Form. *Annals of the Association of American Geographers* 84: 557-584.

Smith, B. D. 2001. Low-Level Food Production. *Journal of Archaeological Research* 9: 1-43.

—. 2007. Niche Construction and the Behavioral Context of Plant and Animal Domestication. *Evolutionary Anthropology* 16: 188-199.

Smith, M. E. 1979. The Aztec Marketing System and Settlement Pattern in the Valley of Mexico: A Central Place Analysis. *American Antiquity* 44: 110-125.

—. 1987a. Archaeology and the Aztec Economy: The Social Scientific Use of Archaeological Data. *Social Science History* 11: 237-259.

Smith, M. E. 1987b. Household possessions and wealth in agrarian states. *J. Anthropol. Arch.* 6: 297.

Smith, M. E. 1993. Houses and the Settlement Hierarchy in Late Posclassic Morelos: A Comparison of Archaeology and Ethnohistory. In R. S. Santley and K. G. Hirth (eds.), *Prehispanic Domestic Units in Western Mesoamerica: Studies of the Household, Compound, and Residence*: 191-206. Boca Raton, CRC Press, Inc.

—. 1994. Economies and Polities in Aztec-Period Morelos. In M. G. Hodge and M. E. Smith (eds.), *Economies and Polities in the Aztec Realm*: 313-348. Albany, Institute for Mesoamerican Studies.

Smith, M. E., and F. F. Berdan. 1992. Archaeology and the Aztec Empire. *World Archaeology* 23: 353-367.

Smith, M. E., and C. Heath-Smith. 2000. Rural Economy in Late Postclassic Morelos: An Archaeological Study. In M. E. Smith and M. A. Masson (eds.), *The Ancient Civilizations of Mesoamerica: A Reader*: 217-235. Malden, Blackwell Publishers Ltd.

Smith, M. E., and K. G. Hirth. 1988. The Development of Prehispanic Cotton-Spinning Technology in Western Morelos, Mexico. *Journal of Field Archaeology* 15: 349-258.

Smith, M. E., and M. G. Hodge. 1994. An Introduction to Late Postclassic Economies and Polities. In M. G. Hodge and M. E. Smith (eds.), *In Economies and Polities in the Aztec Realm*: 1-42. Austin, University of Texas Press.

Smyth, M. P. 1989. Domestic Storage Behavior in Mesoamerica: An Ethnoarchaeological Approach. In S. Michael B. (ed.) *Archaeological Method and Theory* vol. 1: 89-138. Tucson, University of Arizona Press.

Solís, E. C. 1992. *Anales de Tecamachalco. 1398-1590*. Puebla, Gobierno del Estado de Puebla.

Stadleman, R. 1940. *Maize Cultivation in Northwestern Guatemala*. Washington, D.C., Carnegie Institution Contributions to American Anthropology and History, v. 16.

Stahle, D. W., and M. K. Cleveland. 1994. Tree-Ring Reconstructed Rainfall over the Southeastern U.S.A. During the Medieval Warm Period and Little Ice Age. *Climatic Change* 26.

Stahle, D. W., J. Villanueva, D. J. Burnette, J. Cerano, R. Heim Jr., F. K. Fye, R. A. Soto, M. D. Therrell, M. K. Cleveland, and D. K. Stahle. 2011. Major Mesoamerican Droughts of the Past Millennium. *Geophysical Research Letters* 38(L05703): 1-4.

Steggerda, M. 1941. *Maya Indians of Yucatán*. Washington, D.C., Carnegie Institution of Washington Publication 531.

Stein, G. 2004. Economic Organization of North Mesopotamian Urbanism. In G. M. Feinman and L. M. Nicholas (eds.), *Archaeological Perspectives on Political Economies*: 61-78. Salt Lake City, The University of Utah Press.

Stone, G. D. 1996. *Settlement Ecology: The Social and Spatial Organization of Kofyar Agriculture*. Tucson, The University of Arizona Press.

Sullivan, T. D. 1987. *Documentos Tlaxcaltecas del Siglo XVI en lengua Náhuatl: introducción, paleografía, traducción y notas*. México, D.F., Universidad Nacional Autónoma de México.

Svečnjak, Z., B. Varga, and J. Butorac. 2006. Yield Components of Apical and Subapical Ear Contributing to the Grain Yield Responses of Prolific Maize at High and Low Plant Populations. *Journal of Agronomy & Crop Science* 192: 37-42.

Terán, S., and C. H. Rasmussen. 1994. *La milpa de los Mayas: La agricultura de los Mayas prehispánicos y *actuales en el noreste de Yucatán*, Terán and Rasmussen/ DANIDA.

Therrell, M. D. 2005. Tree rings and 'El Año del Hambre' in Mexico. *Dendrochronologia* 22: 203-207.

Therrell, M. D., D. W. Stahle, J. Villanueva Diaz, E. H. Cornejo, and M. K. Cleaveland. 2006. Tree-Ring Reconstructed Maize Yield in Central Mexico: 1474-2001. *Climatic Change* 74: 493-504.

Thompson, G. D., and P. N. Wilson. 1994. Common Property as an Institutional Response to Environmental Variability. *Contemporary Economic Policy* 12: 10-21.

Thomson, H. 1986. Subsistence Agriculture in Papua New Guinea. *Journal of Rural Studies* 2: 233-243.

Torquemada, J. d. 1969 [1615]. *Monarquía Indiana. Tomo 2*. México, D.F., Editorial Porrúa.

Torres, B. 1985. Las plantas útiles en el México antiguo según las fuentes del siglo XVI. In T. Rojas and W. T. Sanders (eds.), *Historia de la agricultura Época prehispánica siglo XVI*, vol. I: 53-128. México, D.F., Instituto Nacional de Antropología e Historia.

Trautmann, W. 1997. Examen del proceso de despoblamiento en Tlaxcala durante la Época Colonial. In Á. García Cook and B. L. Merino (eds.), *Antología de Tlaxcala*, vol. II: 51-56. México D.F., Instituto Nacional de Antropología e Historia.

Trigger, B. G. 2003. *Understanding Early Civilizations: A Comparative Study*. Cambridge, Cambridge University Press.

Urrutia, V. 1967. *Corn Production and Soil Fertility Under Shifting Cultivation in Uaxactun, Guatemala*. Unpublished M.A. thesis, Department of Agriculture. Gainesville, University of Florida.

Utts, J. M. 2005. *Seeing Through Statistics*. Belmont, Thomson Brooks/Cole.

Valle, P. 1992. *Memorial de los indios de Tepetlaoztoc o Códice Kingsborough "... A cuatrocientos cuarenta años"*. México, D.F., Instituto Nacional de Antropología e Historia.

Vandermeer, J. H. 1989. *The ecology of intercropping*. New York, Cambridge University Press.

Velazquez, P. F. 1945. *Codice Chimalpopoca: anales de Cuauhtitlan y leyenda de los soles*. México, D.F., Universidad Nacional Autónoma de México.

Villa, A. 1945. *The Maya of East-Central Quintana Roo*. Washington, D.C., Carnegie Institution.

Webster, D. L. 1988. Household Remains of the Humblest Maya. *Journal of Field Archaeology* 15: 169-190.

Wellhausen, E. J., L. M. Roberts, and E. Hernandez X., in collaboration with P. C. Mangelsdorf. 1952. *Races of Maize in Mexico: Their Origin, Characteristics and Distribution* Harvard, Bussey Institution of Harvard University.

Werner, G. 2012. *Los suelos en el Estado de Tlaxcala: Homenaje a Gerd Werner*. Tlaxcala, Gobierno del Estado de Tlaxcala.

Werner, G., G. Miehlich, and H. Aeppli. 1978. Los Suelos de la Cuenca Alta de Puebla-Tlaxcala y sus alrededores. *comunicaciones* Suplemento VI.

Whitmore, T. M., and B. L. Turner II. 1992. Landscapes of Cultivation in Mesoamerica on the Eve of the Conquest. *Annals of the Association of American Geographers* 82: 402-425.

—. 2001. *Cultivated Landscapes of Middle America on the Eve of the Conquest*. New York, Oxford University Press.

Wilk, R. R. 1982. *Agricultural Ecology and Domestic Organization Among the Kekchi Maya*. Unpublished Ph.D. dissertation, Department of Anthropology. Tucson, University of Arizona.

—. 1991. *Household Ecology: Economic Change and Domestic Life Among the Kekchi Maya in Belize*. Tucson, The University of Arizona Press.

Wilken, G. C. 1987. *Good Farmers: Traditional Agricultural Resource Management in Mexico and Central America*. Berkeley, University of California Press.

Wilshusen, R. H., and G. D. Stone. 1990. Soils in Early Agriculture. *World Archaeology* 22: 104-114.

Williams, B. J., and H. R. Harvey. 1997. *The Códice de Santa María Asunción: Households and Lands in Sixteenth-Century Tepetlaoztoc*. Salt Lake City, University of Utah Press.

Williams, B. J., and M. d. C. Jorge y Jorge. 2008. Aztec Arithmetic Revisited: Land-Area Algorithms and Acolhua Congruence Arithmetic. *Science* 320: 72-77.

Willis, M. G., and D. J. Horne. 1992. Soil water repellency. *Advances in Soil Science* 20: 91-146.

Wolf, E. R. 1966. *Peasants*. New Jersey, Prentice-Hall.

Xelhuantzi, M. S. 1994. Estudio palinológico de cuatro sitios ubicados en la cuenca de Zacapu: fondo de la ciénega, contacto Lomas-ciénega, pantano interno y Loma Alta. In P. Pétrequin (ed.) *8000 Años de la Cuenca de Zacapu: evolución de los paisajes y primeros desmontes*: 77-80. México, D.F., Centre D´Études Mexicaines et Centraméricaines México.

Yamamoto, F. 1992. Effects of Depth of Flooding on Growth and Anatomy of Stems and Knee Roots of Taxodium Distichum. *IAWA Bulletin* 13: 93-104.

Yemelyanau, M. 2009. Second Agriculture in Belarus and Ukraine: Subsistence or Leisure? *Working Paper Series, Belarusian Economic Research and Outreach Center* 008: 4-27.

Yiriode, E. K., A. S. Langyintuo, and W. Dogbe. 2006. Economics of the Impact of alternative rice cropping systems on subsistence farming: Whole-Farm analysis in northern Ghana. *Agricultural Systems* 91: 102-121.

Yoneda, K. 1991. *Los mapas de Cuauhtinchan y la historia cartográfica prehispánica*. México, D.F., CIESAS.

—. 1994. *Cartografía y linderos en el Mapa de Cuauhtinchan No. 4*. México, D.F., INAH- BUAP.

—. 2005. *Mapa de Cuauhtinchan núm. 2*. México, D.F., CIESAS.

Zier, C. J. 1992. Intensive Raised Field Agriculture in a Posteruption Environment, El Salvador. In T. W. Killion (ed.) *Gardens of Prehistory: the Archaeology of Settlement Agriculture in Greater Mesoamerica*: 217-233. Tuscaloosa, University of Alabama Press.

Zimmerer, K. S. 1991. Wetland Production and Smallholder Persistence: Agricultural Change in a Highland Peruvian Region. *Annals of the Association of American Geographers* 81: 443-463.

—. 1993. Agricultural Biodiversity and Peasant Rights to Subsistence in the Central Andes During the Inca Rule. *Journal of Historical Geography* 19: 15-32.

Zorita, A. d. 1942. *Breve y sumaria relación de los señores de la Nueva España*. México, D.F., Universidad Nacional Autónoma de México.

Appendix
2009 Ethnographic Survey:
Field Registers

Field number	UTM (centroid X)	UTM (centroid Y)	Average volume of the ear (cm3)	Estimated weight of the kernels based on the volume of the ear at 14% humidity (grams)	Estimated sowing density (based on 3 kernels per group of plants)	Estimated number of plants that survived (total per hectare)	Estimated number of plants that produced an ear (total per hectare)	Estimated productivity (kg/ha) at 14% humidity	Estimated volume of shelled maize (cm3)
1	645159	2115889	46.55	38.75	37129	31188	12376	480	677
2	625343	2124295	47.80	39.80	47561	38415	20122	801	1130
3	622679	2124853	48.62	40.48	43333	31667	14444	585	825
4	613412	2124308	58.36	48.59	39655	32759	17241	838	1178
5	613061	2124731	53.57	44.61	34500	28000	15500	691	974
6	614317	2125258	53.10	44.21	39655	29310	13793	610	859
7	621178	2122990	46.64	38.83	39063	30729	15625	607	857
8	619315	2123988	55.91	46.55	48148	38272	20370	948	1335
9	653166	2113064	45.36	37.76	39796	31633	12245	462	653
10	620965	2123815	54.19	45.12	45732	34756	18293	825	1162
11	625970	2122777	60.66	50.51	44118	31176	16471	832	1170
12	626527	2122693	59.09	49.20	42857	30952	17262	849	1194
13	633451	2118138	56.93	47.40	50000	36420	20370	966	1359
14	613857	2104075	33.08	27.53	42135	30899	8427	232	330
15	646619	2114161	42.94	35.75	37500	27976	11905	426	602
16	640980	2117481	52.54	43.75	40206	31443	17526	767	1080
17	628987	2121274	65.76	54.76	41071	35119	20238	1108	1556
18	622714	2121346	58.13	48.40	45181	33133	16867	816	1148
19	629693	2122424	52.57	43.77	40116	29070	14535	636	896
20	625167	2123429	51.74	43.08	43605	36047	16279	701	988
21	613172	2104521	41.88	34.86	37500	28500	11000	383	543
22	626039	2121731	50.59	42.12	45349	34302	18023	759	1070
23	617527	2123759	55.19	45.95	38415	31707	17073	785	1105
24	624306	2123457	58.08	48.37	43548	34946	19355	936	1317
25	612530	2124325	48.74	40.58	46552	32759	13218	536	757
26	618298	2123108	58.26	48.51	37129	31188	17327	841	1182
27	643751	2116452	58.63	48.82	40741	32716	17284	844	1187
28	615899	2124534	54.89	45.71	44444	35802	16049	734	1033
29	629231	2098066	27.19	22.62	38372	26744	2326	53	75
30	633372	2118707	58.41	48.64	37500	32065	18478	899	1264
31	622975	2122068	51.22	42.65	45181	37349	16867	719	1014
32	628976	2119915	64.89	54.04	44118	32941	17647	954	1339
33	644278	2116460	41.55	34.59	43125	36875	13750	476	673
34	622771	2123627	55.30	46.05	43902	37195	20732	955	1344
35	622389	2121746	51.12	42.56	39759	31325	16265	692	976
36	622546	2122775	52.77	43.94	48148	35802	19136	841	1185
37	644760	2115736	42.28	35.20	40099	31683	12376	436	616
38	614928	2124810	53.70	44.71	45882	33529	17647	789	1111
39	637891	2117446	61.87	51.52	41379	31609	17816	918	1290
40	647439	2114057	57.70	48.05	39394	29798	15657	752	1058
41	612400	2117069	48.80	40.63	40761	29891	13043	530	748

Field number	UTM (centroid X)	UTM (centroid Y)	Average volume of the ear (cm3)	Estimated weight of the kernels based on the volume of the ear at 14% humidity (grams)	Estimated sowing density (based on 3 kernels per group of plants)	Estimated number of plants that survived (total per hectare)	Estimated number of plants that produced an ear (total per hectare)	Estimated productivity (kg/ha) at 14% humidity	Estimated volume of shelled maize (cm3)
42	598782	2110063	42.21	35.14	38265	32143	15306	538	761
43	601813	2112054	39.53	32.90	41772	30380	10759	354	502
44	617152	2103177	36.50	30.38	46988	35542	10843	329	468
45	622513	2115816	53.31	44.39	39894	32979	18617	826	1164
46	607458	2118751	31.72	26.40	35204	28061	6633	175	249
47	615666	2122230	59.35	49.42	40761	33152	17935	886	1246
48	607898	2115139	49.77	41.44	40500	33000	16000	663	935
49	617139	2121768	50.20	41.80	48214	39881	19643	821	1158
50	616023	2119692	65.33	54.40	44828	31609	17816	969	1361
51	623340	2115528	46.90	39.05	40323	30108	15591	609	860
52	613605	2112294	39.39	32.79	37500	29891	8152	267	379
53	609019	2114476	29.51	24.56	49367	41772	8228	202	288
54	609355	2112083	34.61	28.81	40909	32323	6061	175	248
55	612260	2111587	46.86	39.01	40909	30114	9659	377	532
56	609975	2116923	32.25	26.84	48795	36747	9036	243	345
57	604681	2118335	55.57	46.27	39286	32143	17262	799	1124
58	630204	2117075	51.09	42.54	40116	33721	18023	767	1081
59	598810	2112316	33.98	28.28	33871	27957	6452	182	259
60	642429	2115095	47.32	39.40	36316	31053	14737	581	820
61	615968	2123717	60.08	50.03	37097	26882	13441	672	945
62	601021	2114464	34.31	28.56	46429	39286	9524	272	387
63	635245	2098533	30.33	25.24	42857	34524	2976	75	107
64	640083	2112306	58.19	48.46	38372	29070	15698	761	1070
65	615705	2121840	47.63	39.65	43103	36207	20115	798	1126
66	626921	2119665	54.30	45.22	41209	30769	14286	646	910
67	607110	2114549	49.51	41.22	46296	33333	15432	636	897
68	638595	2116157	59.09	49.20	43820	34831	20225	995	1399
69	628229	2113214	40.15	33.42	39130	31522	10326	345	489
70	605590	2113661	31.91	26.56	43373	33133	7831	208	296
71	621499	2117443	53.02	44.15	35567	26804	14433	637	898
72	635435	2116836	60.56	50.43	41489	29255	13830	697	980
73	602350	2113629	50.10	41.71	37097	28495	13441	561	791
74	628133	2095564	28.12	23.40	41250	30000	3125	73	105
75	610707	2120506	47.41	39.47	43671	36076	16456	649	917
76	628497	2115351	40.65	33.84	42593	30247	9259	313	444
77	612743	2116357	49.87	41.53	39205	27273	15341	637	898
78	595915	2112570	44.80	37.30	44643	35119	14881	555	784
79	611387	2116869	42.99	35.79	39796	30102	15306	548	775
80	630734	2111813	49.31	41.05	38710	27957	11828	486	685
81	634734	2101699	44.67	37.18	41772	34810	13291	494	698
82	610782	2115969	41.88	34.87	41935	30108	11828	412	584
83	598580	2115080	42.25	35.17	40588	29412	11176	393	556
84	621773	2119711	63.38	52.78	39205	32386	18182	960	1348
85	601000	2109687	50.45	42.01	37500	31250	18182	764	1077
86	612608	2114719	52.73	43.90	42857	34066	17582	772	1088
87	608869	2111596	38.36	31.93	37113	28351	10825	346	490
88	632326	2117319	51.52	42.89	40116	29651	13372	574	808

Appendix

Field number	UTM (centroid X)	UTM (centroid Y)	Average volume of the ear (cm3)	Estimated weight of the kernels based on the volume of the ear at 14% humidity (grams)	Estimated sowing density (based on 3 kernels per group of plants)	Estimated number of plants that survived (total per hectare)	Estimated number of plants that produced an ear (total per hectare)	Estimated productivity (kg/ha) at 14% humidity	Estimated volume of shelled maize (cm3)
89	609041	2119136	40.86	34.01	42135	32022	9551	325	460
90	625701	2114079	57.62	47.98	37879	31818	17172	824	1159
91	608207	2119701	35.24	29.33	39205	33523	9659	283	402
92	624429	2109300	38.03	31.65	42614	32386	8523	270	383
93	606920	2112265	54.06	45.02	38333	31667	17778	800	1127
94	638065	2102934	26.68	22.20	50000	40741	4938	110	157
95	611847	2112751	41.00	34.13	41071	29762	8929	305	431
96	606512	2111940	57.15	47.59	36264	30769	16484	784	1104
97	630289	2115467	26.63	22.16	40500	33500	1500	33	48
98	610919	2113518	34.56	28.76	38764	30899	6742	194	276
99	606648	2118497	56.08	46.69	36702	30319	16489	770	1084
100	641202	2114690	57.19	47.62	34500	26000	12500	595	838
101	598090	2116686	41.02	34.14	41753	30412	11340	387	548
102	642627	2113004	45.24	37.67	47647	38824	14118	532	751
103	616184	2115054	54.67	45.52	38333	29444	15556	708	997
104	629755	2112639	47.71	39.72	42857	32738	15476	615	868
105	631772	2118054	61.98	51.61	32474	23196	12887	665	935
106	596786	2114505	48.84	40.67	38333	32222	16667	678	956
107	632882	2115165	59.48	49.53	32813	24479	12500	619	871
108	608049	2118212	46.48	38.70	43085	35106	13298	515	727
109	630321	2118039	57.77	48.11	42188	34896	19792	952	1339
110	606652	2116265	34.22	28.48	45000	35556	8333	237	337
111	610042	2115708	49.72	41.40	41053	30000	16316	675	953
112	637336	2096159	29.56	24.60	36316	30000	3684	91	129
113	643320	2112065	50.95	42.42	39000	29500	15500	658	927
114	605534	2116540	34.32	28.57	34848	25253	6566	188	267
115	625773	2113267	56.28	46.86	44828	37356	20690	970	1365
116	622908	2118359	53.04	44.16	39474	31579	16316	721	1015
117	629140	2102667	31.57	26.27	35204	26020	5102	134	191
118	618638	2121195	53.48	44.53	37879	29798	13636	607	855
119	627303	2118089	48.74	40.58	46552	32759	17816	723	1020
120	604571	2111337	53.51	44.56	40323	28495	11828	527	742
121	620635	2121588	53.18	44.28	33333	25758	13131	581	819
122	618661	2099009	30.73	25.57	44444	32716	7407	189	270
123	599999	2113335	33.60	27.97	43125	31250	8125	227	323
124	612944	2113427	47.57	39.60	40323	32796	13978	554	781
125	606012	2118941	36.10	30.04	39000	32000	7500	225	320
126	598611	2115791	57.67	48.02	37079	28090	16292	782	1101
127	643701	2110585	53.02	44.14	41053	34737	16316	720	1015
128	632754	2116455	61.49	51.21	37500	32065	15761	807	1134
129	622879	2116737	47.32	39.39	38710	30645	17204	678	957
130	610590	2118282	49.15	40.93	41209	29670	13187	540	761
131	603490	2111469	40.86	34.01	40909	32955	10795	367	520
132	621158	2120434	54.81	45.64	43125	35000	18750	856	1205
133	625479	2112125	32.70	27.22	41209	34066	7143	194	277
134	630406	2113146	53.65	44.67	37129	29208	13861	619	872
135	632735	2116730	55.34	46.08	46988	37952	19880	916	1290

Field number	UTM (centroid X)	UTM (centroid Y)	Average volume of the ear (cm3)	Estimated weight of the kernels based on the volume of the ear at 14% humidity (grams)	Estimated sowing density (based on 3 kernels per group of plants)	Estimated number of plants that survived (total per hectare)	Estimated number of plants that produced an ear (total per hectare)	Estimated productivity (kg/ha) at 14% humidity	Estimated volume of shelled maize (cm3)
136	601855	2115609	31.54	26.25	49367	39241	6962	183	260
137	595878	2112827	38.40	31.97	37500	30500	9000	288	408
138	628947	2119065	52.64	43.83	43605	37209	16860	739	1041
139	622182	2110157	34.43	28.66	39894	34043	8511	244	347
140	611301	2122644	46.04	38.33	37912	26923	12088	463	654
141	612437	2122984	50.44	42.00	39375	30000	16875	709	999
142	603094	2116750	40.00	33.30	45000	37222	13889	462	655
143	630842	2113075	44.68	37.20	49367	36709	13291	494	699
144	607629	2111706	35.56	29.60	45506	38764	10674	316	449
145	603545	2114058	37.45	31.17	40116	33140	10465	326	463
146	614324	2122283	58.36	48.59	37059	29412	17059	829	1166
147	619575	2121155	55.24	46.00	36364	27778	16162	743	1047
148	639401	2112083	58.04	48.33	46875	35000	20000	967	1360
149	624099	2118294	49.39	41.12	46552	38506	17816	733	1033
150	615350	2119928	64.91	54.05	45349	31977	18023	974	1368
151	625096	2120207	55.51	46.22	42073	32317	18293	845	1190
152	606315	2118809	33.98	28.28	41071	33333	8929	253	359
153	616520	2114346	36.90	30.72	35938	26563	8333	256	363
154	619502	2120280	58.78	48.94	39205	28409	16477	806	1134
155	613192	2115615	42.30	35.21	42073	35976	14634	515	729
156	602032	2116267	35.05	29.17	41327	29592	6633	193	275
157	627990	2112193	34.81	28.97	45181	32530	9639	279	397
158	639113	2114101	45.57	37.93	48148	38272	17901	679	959
159	608598	2119581	39.01	32.47	35000	28333	9444	307	435
160	613654	2112612	39.98	33.28	35484	30108	11828	394	558
161	627135	2117239	64.89	54.04	37952	31325	18072	977	1371
162	642139	2114354	58.35	48.59	47468	33544	18987	923	1298
163	612638	2121864	33.20	27.63	38764	28090	6180	171	243
164	607640	2113235	35.71	29.72	44022	31522	7065	210	298
165	599290	2111291	49.34	41.08	44828	34483	13793	567	799
166	628640	2118378	52.63	43.82	41209	34066	15934	698	984
167	597951	2112317	37.36	31.09	45882	36471	7647	238	337
168	636616	2117711	69.68	58.03	38764	28090	15730	913	1281
169	649211	2111515	40.02	33.31	40625	33333	11979	399	565
170	644548	2111882	42.97	35.77	43605	34302	12209	437	618
171	637966	2114783	55.98	46.61	41053	33158	18421	859	1209
172	631094	2115628	35.50	29.55	48214	36310	8333	246	350
173	631995	2117696	55.59	46.29	44022	32065	16848	780	1098
174	610805	2121810	49.96	41.59	39063	32292	13021	542	764
175	600408	2116222	44.69	37.20	44318	34091	14205	528	747
176	625693	2114990	60.29	50.20	35204	24490	13776	692	972
177	604504	2112426	40.35	33.59	36364	28283	12121	407	577
178	606259	2116461	32.77	27.27	39655	31609	8046	219	312
179	632304	2106359	31.58	26.28	39375	29375	5625	148	211
180	620791	2119653	54.70	45.55	40909	29293	15657	713	1004
181	611041	2116827	41.74	34.75	43373	32530	14458	502	711
182	627274	2119582	59.80	49.80	40099	30198	15842	789	1109

Field number	UTM (centroid X)	UTM (centroid Y)	Average volume of the ear (cm3)	Estimated weight of the kernels based on the volume of the ear at 14% humidity (grams)	Estimated sowing density (based on 3 kernels per group of plants)	Estimated number of plants that survived (total per hectare)	Estimated number of plants that produced an ear (total per hectare)	Estimated productivity (kg/ha) at 14% humidity	Estimated volume of shelled maize (cm3)
183	647771	2114261	40.32	33.56	45732	34756	14024	471	667
184	608254	2111604	34.44	28.66	43333	32778	6111	175	249
185	631295	2109102	27.83	23.15	39655	29885	3448	80	114
186	601866	2107554	45.95	38.25	42614	35227	13636	522	737
187	605248	2107067	42.35	35.25	39205	33523	10795	381	538
188	640995	2111535	53.83	44.82	39894	29787	15957	715	1007
189	614588	2110076	36.29	30.21	40099	32178	10396	314	446
190	625958	2110064	34.29	28.54	43671	36076	9494	271	385
191	613062	2111197	37.02	30.81	41379	29310	8046	248	352
192	638167	2111125	50.48	42.03	44118	37059	20588	865	1220
193	598152	2109497	43.29	36.04	37113	27835	13402	483	683
194	600858	2110828	52.31	43.56	41935	30645	13978	609	858
195	619525	2109115	43.53	36.24	37879	29293	10606	384	543
196	610627	2110919	36.31	30.22	40244	29268	6707	203	288
197	631029	2106293	29.89	24.87	41209	31319	6593	164	234
198	620935	2111251	43.40	36.13	37129	30198	11881	429	607
199	604092	2107716	34.44	28.66	40449	34270	12921	370	526
200	598365	2107734	40.18	33.44	43671	36076	11392	381	540
201	599285	2107714	52.20	43.47	36628	26744	13372	581	819
202	613261	2110775	34.24	28.50	35106	28191	7447	212	302
203	647226	2111003	45.00	37.46	45349	38372	15116	566	800
204	599511	2109604	49.93	41.57	43605	32558	13953	580	818
205	623075	2107199	34.60	28.80	37500	31500	7500	216	307
206	626056	2109994	40.51	33.72	37952	30120	12651	427	604
207	598836	2105922	50.75	42.26	44118	36471	15882	671	946
208	599424	2108140	42.61	35.48	40449	33708	12360	438	620
209	624604	2110753	35.37	29.44	41772	34810	7595	224	318
210	648311	2111991	50.34	41.91	39394	28283	13636	572	806
211	596778	2110665	47.47	39.52	41053	30000	15263	603	852
212	612047	2109355	30.50	25.38	37931	28736	7471	190	270
213	610719	2108997	40.22	33.48	42073	34146	12195	408	578
214	604730	2110925	53.37	44.44	36702	31383	14362	638	899
215	601827	2109269	55.83	46.49	38710	29032	16129	750	1055
216	598413	2108000	53.41	44.47	34848	26768	13636	606	854
217	613324	2107642	35.40	29.46	37500	31548	9524	281	399
218	606494	2107917	44.44	37.00	42135	35955	16292	603	852
219	632785	2106745	31.19	25.95	45732	32317	3659	95	135
220	607228	2110643	40.88	34.03	44505	37912	15385	524	741
221	613985	2107722	37.62	31.32	37500	26136	7386	231	328
222	608392	2109056	47.44	39.49	33000	26500	14000	553	780
223	602914	2107892	52.68	43.86	38415	28049	11585	508	716
224	645780	2112511	46.75	38.92	35000	27222	12222	476	672
225	609772	2109876	53.12	44.23	39873	29747	17089	756	1065
226	640127	2111344	53.43	44.49	44118	32941	14706	654	921
227	639372	2110501	40.24	33.49	38824	32353	11176	374	530
228	602237	2108475	44.78	37.28	43820	35955	13483	503	710
229	612061	2108523	41.20	34.30	35567	27320	10309	354	501

Field number	UTM (centroid X)	UTM (centroid Y)	Average volume of the ear (cm3)	Estimated weight of the kernels based on the volume of the ear at 14% humidity (grams)	Estimated sowing density (based on 3 kernels per group of plants)	Estimated number of plants that survived (total per hectare)	Estimated number of plants that produced an ear (total per hectare)	Estimated productivity (kg/ha) at 14% humidity	Estimated volume of shelled maize (cm3)
230	597203	2110653	49.55	41.26	40761	33696	15761	650	917
231	613426	2110828	41.68	34.70	40909	31250	11932	414	586
232	605727	2108799	41.48	34.53	40000	33333	13889	480	679
233	596354	2109368	45.33	37.74	37500	30978	12500	472	666
234	616059	2108019	41.34	34.42	41327	29082	9184	316	447
235	597876	2109232	40.82	33.98	44118	33529	13529	460	651
236	626456	2109400	32.70	27.21	50625	41250	11250	306	436
237	644946	2111236	43.78	36.45	42614	30682	11932	435	615
238	622989	2108329	43.57	36.27	34158	26733	8416	305	432
239	638788	2111341	54.78	45.61	45732	37195	20122	918	1292
240	602676	2108971	38.37	31.94	40000	33889	11111	355	503
241	640798	2112825	60.02	49.98	41250	32500	18125	906	1274
242	609855	2107925	48.06	40.01	41250	31875	13125	525	741
243	615833	2109469	39.51	32.89	40625	29167	8333	274	388
244	633596	2108273	27.41	22.81	49367	41139	5063	115	165
245	614203	2104999	41.98	34.95	39000	30000	9500	332	470
246	636967	2107490	31.98	26.62	38660	27835	6186	165	234
247	636168	2107208	29.59	24.62	42614	35227	3977	98	140
248	642647	2102524	29.66	24.68	37500	28409	6250	154	220
249	623559	2104563	37.33	31.07	40588	32353	8235	256	363
250	626093	2106582	29.84	24.83	44444	37654	8642	215	306
251	634672	2109307	27.69	23.04	50000	41975	6173	142	203
252	606917	2106425	40.93	34.07	39474	28421	12105	412	584
253	620191	2106701	40.53	33.74	36316	27368	11053	373	528
254	637180	2103885	33.21	27.64	38889	32099	6173	171	243
255	629898	2105209	25.23	20.99	40500	33000	3000	63	90
256	618477	2104237	40.49	33.71	33511	27128	8511	287	406
257	649003	2109405	44.93	37.40	48750	35000	13750	514	727
258	603554	2105761	60.92	50.73	35000	26111	13889	705	990
259	627651	2106197	43.56	36.26	35938	26042	10938	397	561
260	628510	2107744	30.13	25.07	49390	39634	9756	245	349
261	629159	2106923	41.05	34.17	31500	25500	7000	239	339
262	643168	2108898	42.27	35.18	33000	25500	11000	387	548
263	637118	2102151	34.66	28.85	42073	34756	8537	246	350
264	635253	2109691	33.99	28.29	44643	33333	5952	168	239
265	634119	2105995	29.57	24.60	35204	29082	6633	163	233
266	608451	2105712	44.09	36.70	43085	35106	14362	527	745
267	647899	2108905	40.61	33.81	39130	28804	10870	367	520
268	641267	2102536	27.83	23.15	40909	32955	3409	79	113
269	632482	2104105	31.23	25.99	45732	34146	6707	174	248
270	621562	2105233	40.80	33.96	41209	32418	8242	280	396
271	637884	2103754	31.65	26.34	37912	27473	6593	174	247
272	607974	2105648	39.49	32.87	46023	32386	10795	355	503
273	640944	2102672	30.40	25.30	39063	30208	6771	171	244
274	622108	2104061	38.99	32.46	36702	28191	10638	345	489
275	644694	2108528	44.34	36.92	34848	26768	9596	354	501
276	625266	2107099	42.84	35.66	43820	33146	12360	441	623

APPENDIX

Field number	UTM (centroid X)	UTM (centroid Y)	Average volume of the ear (cm3)	Estimated weight of the kernels based on the volume of the ear at 14% humidity (grams)	Estimated sowing density (based on 3 kernels per group of plants)	Estimated number of plants that survived (total per hectare)	Estimated number of plants that produced an ear (total per hectare)	Estimated productivity (kg/ha) at 14% humidity	Estimated volume of shelled maize (cm3)
277	639599	2103946	29.32	24.40	34737	26316	4211	103	147
278	606153	2106055	46.30	38.55	41071	29762	15476	597	842
279	627158	2103277	25.54	21.25	37097	27957	2151	46	66
280	638683	2106787	30.89	25.70	42391	32065	7609	196	279
281	610256	2106378	41.40	34.46	50625	41250	15625	538	762
282	639172	2104108	33.05	27.51	41209	30769	6044	166	237
283	643405	2108370	44.54	37.08	40741	33951	13580	504	712
284	621023	2105213	46.87	39.02	31188	22772	7921	309	436
285	622668	2103705	33.80	28.13	42391	32065	7609	214	304
286	624942	2104014	37.92	31.56	45732	32317	9146	289	409
287	610188	2107346	43.85	36.51	43548	36559	13978	510	721
288	632064	2105150	30.63	25.49	45181	37952	3614	92	131
289	626479	2105569	38.81	32.31	37129	29703	9901	320	453
290	630776	2103950	28.20	23.46	37912	27473	3297	77	111
291	617430	2104718	40.23	33.49	35393	29213	10112	339	480
292	650583	2111640	46.73	38.91	37500	30978	13043	507	717
293	619630	2106815	41.00	34.13	41327	29592	8673	296	419
294	641994	2110773	45.77	38.10	42135	35955	14607	557	786
295	648737	2110844	50.39	41.96	35106	28191	13298	558	787
296	626281	2108646	38.46	32.02	42593	34568	10494	336	476
297	638652	2105998	30.59	25.46	44643	31548	6548	167	238
298	632048	2107746	34.79	28.96	39130	30435	6522	189	268
299	621561	2107140	38.31	31.89	39394	29798	11616	370	525
300	611098	2105528	47.19	39.29	40244	29878	12195	479	676
301	646869	2108460	49.13	40.91	43902	34756	18293	748	1056
302	625395	2105951	35.89	29.87	38614	29208	8416	251	357
303	641306	2110054	43.24	36.00	44022	30978	12500	450	636
304	637463	2106754	33.84	28.16	36702	30851	6383	180	256
305	618912	2104643	46.28	38.53	36364	27273	9596	370	522
306	636405	2108224	27.64	22.99	37500	31771	6250	144	206
307	634295	2105554	34.58	28.78	38660	30412	5670	163	232
308	612199	2105107	38.95	32.42	42073	35366	12805	415	588
309	637530	2107330	31.54	26.24	36667	25556	6667	175	249
310	641507	2109589	38.01	31.64	39474	33158	11053	350	496
311	624337	2105379	32.97	27.44	45000	36250	10000	274	390
312	609567	2106360	44.12	36.73	47468	33544	13924	511	723
313	631778	2104678	25.69	21.37	39375	28750	3125	67	96
314	640690	2099995	37.66	31.35	39796	30102	5612	176	250
315	634429	2103553	28.80	23.96	37912	31319	7143	171	245
316	620443	2104405	33.68	28.03	43605	32558	8721	244	348
317	624150	2108096	44.58	37.11	43548	32796	11828	439	620
318	646272	2108669	40.56	33.76	40323	32796	12366	417	591
319	604104	2105707	40.88	34.03	32673	24257	9901	337	477
320	633828	2099350	24.07	20.02	38298	28191	2128	43	61
321	636837	2101040	30.45	25.34	38660	30928	5155	131	186
322	622538	2101277	36.46	30.35	34158	24752	6931	210	299
323	610662	2104399	45.79	38.13	40206	31959	17526	668	944

Field number	UTM (centroid X)	UTM (centroid Y)	Average volume of the ear (cm3)	Estimated weight of the kernels based on the volume of the ear at 14% humidity (grams)	Estimated sowing density (based on 3 kernels per group of plants)	Estimated number of plants that survived (total per hectare)	Estimated number of plants that produced an ear (total per hectare)	Estimated productivity (kg/ha) at 14% humidity	Estimated volume of shelled maize (cm3)
324	612800	2103288	51.70	43.05	32474	23196	10309	444	625
325	638889	2099941	22.74	18.91	38764	32022	1685	32	46
326	641227	2101083	33.52	27.90	42188	30729	6250	174	248
327	639950	2097385	33.59	27.96	43671	32278	4430	124	176
328	639384	2100753	27.33	22.74	48750	34375	4375	99	142
329	623616	2101463	42.91	35.72	38333	27222	12222	437	618
330	609858	2104030	48.93	40.74	42188	34375	16667	679	958
331	631992	2099134	23.36	19.43	34158	25248	2475	48	69
332	619324	2102457	34.35	28.59	43820	30899	7865	225	320
333	620670	2103299	35.33	29.40	34158	28713	9901	291	414
334	636873	2097887	30.58	25.44	43671	36076	3797	97	138
335	630247	2099770	27.70	23.05	35567	27835	2577	59	85
336	630989	2100670	29.39	24.46	38614	27228	3960	97	138
337	623712	2103396	32.74	27.25	42391	32065	7609	207	295
338	637081	2098767	28.81	23.97	33871	28495	2151	52	74
339	636742	2099746	28.52	23.73	40625	28646	4688	111	159
340	627752	2102551	29.25	24.34	39063	28646	3646	89	127
341	636549	2098314	24.41	20.30	38889	32099	3704	75	108
342	620383	2102805	36.29	30.21	39759	33735	12651	382	543
343	627445	2101462	30.86	25.68	42188	34896	6250	161	229
344	625258	2100365	25.01	20.80	43605	32558	2907	60	87
345	635529	2100186	29.92	24.90	40116	33140	3488	87	124
346	623380	2102470	44.46	37.02	36264	29121	10989	407	575
347	625934	2101714	25.19	20.96	40500	31500	2000	42	60
348	639164	2100106	32.45	27.00	37059	27059	5882	159	226
349	626539	2100973	26.79	22.29	41667	35000	5556	124	177
350	637931	2098569	28.76	23.93	37500	30978	4891	117	167
351	640378	2098576	28.74	23.91	32474	24742	4124	99	141
352	626886	2099907	25.26	21.02	37931	30460	3448	72	104
353	636818	2100581	25.63	21.32	46296	32716	2469	53	75
354	615855	2102437	37.36	31.10	48148	38272	11111	346	490
355	626595	2097301	28.87	24.02	37097	29570	4301	103	148
356	615737	2095994	30.79	25.62	37931	27586	2299	59	84
357	620528	2098155	39.72	33.06	39063	28646	8854	293	415
358	622150	2098793	32.97	27.44	40116	33140	7558	207	295
359	612244	2095487	27.00	22.46	40909	33333	3030	68	97
360	605224	2104246	37.44	31.16	47468	35443	12025	375	532
361	605710	2096055	26.95	22.42	44118	36471	4118	92	132
362	617723	2099761	37.97	31.61	43548	34409	11290	357	506
363	612731	2106886	41.25	34.34	33158	27368	7895	271	384
364	613504	2101147	44.92	37.39	38764	30899	8989	336	475
365	614950	2099701	34.17	28.44	34615	26374	7143	203	289
366	609657	2102152	40.52	33.73	38415	30488	9756	329	466
367	610352	2101903	40.07	33.35	37879	28283	8081	270	382
368	605722	2104351	36.65	30.51	47468	36709	12658	386	548
369	603419	2104946	47.09	39.20	43125	31250	13125	515	726
370	611985	2101477	31.96	26.60	39560	30220	7692	205	291

Appendix

Field number	UTM (centroid X)	UTM (centroid Y)	Average volume of the ear (cm3)	Estimated weight of the kernels based on the volume of the ear at 14% humidity (grams)	Estimated sowing density (based on 3 kernels per group of plants)	Estimated number of plants that survived (total per hectare)	Estimated number of plants that produced an ear (total per hectare)	Estimated productivity (kg/ha) at 14% humidity	Estimated volume of shelled maize (cm3)
371	613556	2095026	28.79	23.96	50000	42593	6173	148	211
372	612680	2096863	40.33	33.57	48795	41566	12651	425	601
373	633924	2097012	29.33	24.41	49390	38415	4268	104	149
374	617560	2100735	33.76	28.10	47647	36471	11765	331	470
375	609383	2093752	30.55	25.42	47647	36471	5882	150	213
376	610938	2096439	30.17	25.10	46429	38690	5952	149	213
377	602188	2111244	42.61	35.47	42632	34737	13684	485	687
378	608159	2104036	42.10	35.05	41667	29444	13333	467	661
379	614085	2100591	40.38	33.62	43103	36207	11494	386	547
380	601271	2104059	59.08	49.20	35870	26087	15217	749	1053
381	633349	2102522	28.35	23.59	45181	33735	3012	71	102
382	613819	2093670	27.98	23.28	37500	31250	3977	93	132
383	619428	2097513	40.17	33.44	36264	25824	9341	312	442
384	617661	2096287	29.28	24.37	45570	31646	3797	93	132
385	630879	2098287	26.40	21.96	38764	30899	2247	49	71
386	614067	2097030	43.85	36.50	35393	30337	10674	390	551
387	605358	2095100	32.60	27.13	38660	29381	3093	84	119
388	613213	2096344	34.54	28.74	45506	36517	10112	291	413
389	628998	2097588	29.43	24.49	36364	25758	4545	111	159
390	612851	2097379	43.83	36.49	38710	30108	10753	392	555
391	614247	2096308	35.22	29.31	34158	23762	7426	218	309
392	632350	2096334	25.63	21.32	38614	29208	1980	42	61
393	610393	2102616	35.64	29.66	43103	32759	9770	290	412
394	615035	2103195	34.35	28.59	49390	34756	7317	209	297
395	629622	2100935	28.05	23.34	38889	28395	3086	72	103
396	603921	2105243	38.48	32.03	45000	32222	11667	374	530
397	617530	2097865	43.74	36.41	31818	22727	7576	276	390
398	607570	2104597	55.81	46.47	39560	30220	15385	715	1006
399	626595	2097044	29.50	24.54	43902	35366	5488	135	192
400	608205	2096057	26.41	21.97	41379	31609	3448	76	109
401	622751	2099238	31.75	26.42	45349	34302	6977	184	263
402	604654	2096755	27.74	23.08	42073	31098	2439	56	81
403	617921	2101547	39.32	32.73	43103	30460	8046	263	373
404	610441	2103421	51.36	42.77	43820	35393	15730	673	948
405	621984	2097096	39.25	32.67	41489	34043	10106	330	468
406	622400	2098765	31.94	26.58	40500	31000	7500	199	284
407	621572	2100263	31.62	26.32	38764	33146	10674	281	400
408	618278	2098355	42.36	35.26	49367	35443	10127	357	505
409	621285	2099929	32.30	26.88	48795	38554	7831	210	300
410	630391	2098212	24.35	20.25	40323	31720	1613	33	47
411	619644	2096736	41.61	34.64	39205	31250	11932	413	585
412	613115	2102706	34.70	28.88	41489	32447	7979	230	328
413	612089	2102919	37.94	31.58	42857	34524	5952	188	267
414	623961	2096340	28.53	23.74	37895	26316	3158	75	107
415	616343	2100054	34.06	28.35	45570	38608	10127	287	408
416	618136	2096289	28.33	23.58	44118	33529	4706	111	159
417	609354	2103697	40.55	33.75	34848	24242	10101	341	483

Field number	UTM (centroid X)	UTM (centroid Y)	Average volume of the ear (cm3)	Estimated weight of the kernels based on the volume of the ear at 14% humidity (grams)	Estimated sowing density (based on 3 kernels per group of plants)	Estimated number of plants that survived (total per hectare)	Estimated number of plants that produced an ear (total per hectare)	Estimated productivity (kg/ha) at 14% humidity	Estimated volume of shelled maize (cm3)
418	616963	2096670	45.37	37.77	33673	26020	8673	328	463
419	614478	2095531	40.25	33.50	41860	34884	13372	448	634
420	606162	2096843	22.92	19.06	39894	32979	2128	41	58
421	605123	2105470	41.80	34.79	41071	34524	13690	476	674
422	617139	2099443	39.70	33.04	40909	32955	9659	319	452
423	601524	2103475	41.70	34.71	48795	39157	15663	544	769
424	624944	2098408	31.19	25.95	33871	24194	5376	140	199
425	619860	2096596	40.43	33.65	44828	32184	11494	387	548
426	630466	2094551	27.60	22.97	47468	35443	1899	44	62
427	611079	2093244	27.91	23.22	40323	31720	4839	112	161
428	618273	2093834	30.87	25.69	34500	24500	3000	77	110
429	610869	2095056	29.74	24.74	42593	32716	4938	122	174
430	610568	2094240	27.87	23.19	37931	31609	4023	93	133
431	619777	2093005	30.07	25.02	34500	29000	6500	163	232
432	630276	2095477	26.14	21.75	46296	35185	4321	94	135
433	615613	2091438	24.91	20.72	48148	33951	4321	90	129
434	617979	2092725	27.10	22.55	40588	31765	3529	80	114
435	615339	2091948	28.31	23.56	41667	35556	4444	105	150
436	617507	2093183	23.07	19.19	37097	29032	1613	31	45
437	606082	2094501	30.50	25.38	40500	31500	4000	102	145
438	617033	2094427	21.45	17.84	46875	36875	1875	33	48
439	618202	2094342	26.56	22.10	43548	33333	2688	59	85
440	623126	2094565	31.06	25.85	42614	30682	5114	132	188
441	616141	2092758	34.47	28.69	37500	27717	5435	156	222
442	621628	2093233	27.21	22.64	44118	31176	4706	107	152
443	616207	2093030	31.68	26.36	33333	28283	5556	146	209
444	619304	2095968	25.44	21.17	43605	30814	2326	49	71
445	630808	2095276	26.87	22.36	43902	33537	4268	95	137
446	611697	2092592	28.09	23.37	34615	28022	4396	103	147
447	616893	2092244	25.30	21.04	39000	28000	1500	32	45
448	609951	2095469	29.90	24.88	45181	33735	4819	120	171
449	621619	2093423	28.95	24.09	32673	25248	2970	72	102
450	631480	2095465	26.06	21.68	47561	37805	5488	119	171
451	624493	2093393	27.79	23.12	43820	36517	5056	117	167
452	612717	2094145	27.12	22.56	38710	31183	3226	73	104
453	611470	2096013	28.36	23.60	35795	26136	3409	80	115
454	622648	2095279	30.29	25.20	43103	36207	4023	101	145
455	616876	2095860	28.19	23.46	40323	30108	1613	38	54
456	613920	2092286	22.07	18.36	38333	27222	1667	31	44
457	621969	2093906	32.73	27.24	33000	25500	5500	150	213
458	607705	2095193	23.48	19.53	37931	30460	1724	34	48
459	613378	2092619	30.18	25.11	40761	33696	4891	123	175
460	631819	2094094	20.87	17.35	41566	31928	2410	42	60
461	634879	2094500	28.59	23.79	42353	34118	5882	140	200
462	630178	2091030	25.61	21.30	43671	34810	2532	54	77
463	616955	2090458	27.98	23.28	45732	34146	4878	114	162
464	633872	2092527	33.06	27.51	40761	28804	4891	135	192

Field number	UTM (centroid X)	UTM (centroid Y)	Average volume of the ear (cm3)	Estimated weight of the kernels based on the volume of the ear at 14% humidity (grams)	Estimated sowing density (based on 3 kernels per group of plants)	Estimated number of plants that survived (total per hectare)	Estimated number of plants that produced an ear (total per hectare)	Estimated productivity (kg/ha) at 14% humidity	Estimated volume of shelled maize (cm3)
465	634966	2094946	32.89	27.38	37113	26804	3608	99	141
466	636291	2092236	28.84	24.00	34848	27273	3535	85	121
467	625269	2093463	25.52	21.23	39655	29885	2299	49	70
468	630020	2088963	29.21	24.31	42135	29775	3933	96	137
469	640299	2096611	30.66	25.51	41489	30851	5319	136	194
470	627497	2089998	28.11	23.39	37500	26500	6000	140	201
471	622908	2090654	28.65	23.84	41860	30233	2326	55	79
472	635003	2092587	29.62	24.65	36316	27368	3684	91	130
473	631300	2090975	33.07	27.52	34737	24737	5263	145	206
474	616329	2091877	23.70	19.71	45349	37209	1744	34	49
475	630137	2092355	22.81	18.97	46296	38272	1852	35	51
476	614973	2090833	24.56	20.43	37931	27586	2299	47	67
477	630538	2092916	31.19	25.95	40741	29630	5556	144	205
478	612626	2090752	26.66	22.18	38298	27128	2128	47	68
479	627450	2091719	30.10	25.04	38333	31111	1667	42	60
480	626032	2089173	31.76	26.43	37500	28261	4348	115	164
481	637533	2093217	26.31	21.89	40244	32927	3049	67	96
482	623695	2090313	25.30	21.04	44118	37059	2353	50	71
483	631619	2092451	27.44	22.83	39000	27500	3000	68	98
484	636057	2092459	29.09	24.21	47561	37805	6707	162	232
485	622606	2092025	26.66	22.18	41071	35119	4167	92	132
486	633752	2094445	21.95	18.26	39873	29747	1899	35	50
487	630625	2091638	27.02	22.48	40741	30247	3086	69	99
488	632259	2092957	24.80	20.63	36364	29798	3535	73	105
489	629139	2089591	24.04	20.00	31818	24242	1515	30	44
490	628077	2093282	32.34	26.91	42073	34756	4878	131	187
491	636811	2094358	28.82	23.98	39894	29787	4255	102	146
492	635061	2094796	30.70	25.55	36316	25263	3158	81	115
493	631903	2091380	32.04	26.66	41071	30952	5952	159	226
494	627421	2092030	25.26	21.01	45181	36145	4217	89	127
495	628217	2088764	26.24	21.83	45181	38554	4217	92	132
496	612933	2089977	30.00	24.97	45506	37079	1685	42	60
497	619107	2088178	24.81	20.64	48148	38272	2469	51	73
498	624071	2089175	26.03	21.66	37097	31183	1613	35	50
499	613653	2089665	33.46	27.85	39894	28723	5851	163	232
500	618259	2089614	28.30	23.54	39286	30357	1786	42	60